STRATEGIC STUDIES INSTITUTE

G000054171

The Strategic Studies Institute (SSI) is part of the U.S. Army War College and is the strategic-level study agent for issues related to national security and military strategy with emphasis on geostrategic analysis.

The mission of SSI is to use independent analysis to conduct strategic studies that develop policy recommendations on:

- Strategy, planning, and policy for joint and combined employment of military forces;

- Regional strategic appraisals;

- The nature of land warfare;

- Matters affecting the Army's future;

- The concepts, philosophy, and theory of strategy; and,

- Other issues of importance to the leadership of the Army.

Studies produced by civilian and military analysts concern topics having strategic implications for the Army, the Department of Defense, and the larger national security community.

In addition to its studies, SSI publishes special reports on topics of special or immediate interest. These include edited proceedings of conferences and topically oriented roundtables, expanded trip reports, and quick-reaction responses to senior Army leaders.

The Institute provides a valuable analytical capability within the Army to address strategic and other issues in support of Army participation in national security policy formulation.

i

Strategic Studies Institute
and
U.S. Army War College Press

A HARD LOOK AT HARD POWER: ASSESSING THE DEFENSE CAPABILITIES OF KEY U.S. ALLIES AND SECURITY PARTNERS

Gary J. Schmitt
Editor

July 2015

*****

Comments pertaining to this report are invited and should be forwarded to: Director, Strategic Studies Institute and U.S. Army War College Press, U.S. Army War College, 47 Ashburn Drive, Carlisle, PA 17013-5010.

*****

*****

The Strategic Studies Institute and U.S. Army War College Press publishes a monthly email newsletter to update the national security community on the research of our analysts, recent and forthcoming publications, and upcoming conferences sponsored by the Institute. Each newsletter also provides a strategic commentary by one of our research analysts. If you are interested in receiving this newsletter, please subscribe on the SSI website at *www.StrategicStudiesInstitute.army.mil/newsletter.*

ISBN 1-58487-684-0

CONTENTS

FOREWORD

Since the end of World War II, the United States has made maintaining a favorable balance of power in Eurasia a core element of its national security strategy. It did so in good measure by maintaining a large conventional military force that was based not only at home, but also in bases spread across Europe and Asia. That strategy was buttressed by developing security ties and alliances with key powers and frontline states. The implicit bargain was that the United States would help keep the peace on their door front if they would provide access from which American forces could operate and, in turn, maintain credible forces themselves to reinforce and support U.S. efforts at keeping the great power peace.

The question raised by this collection of essays is: Is that bargain unraveling? As the following chapters note, since the end of the great power threat posed by the Soviet Union, both the United States and its principal allies have seen fit to cut the size of their forces substantially and, in most cases, slowed efforts at replacing military systems and platforms. The quandary many of America's allies have faced is, on the one hand, reforming their militaries to make them more expeditionary and useful for addressing various security problems—such as piracy, terrorism, and the instability brought about by collapsing regimes. On the other hand, not having the political resources at home to prioritize defense spending in the face of domestic demands and, more recently, faltering economies are also problems that need to be considered. The result is smaller, half-modernized militaries with often significant gaps in key capabilities.

The strategic problem is that, while its allies and partners have shrunk their militaries, so too has the

United States. It no longer retains a military sized to handle multiple major contingencies at once as it once did and is now facing the prospect of not only continuing to deal with large-scale disorder within the Middle East but also the problematic behavior of two major military powers, China and Russia. In short, at a time when the United States needs the most help, the prospects for receiving it, with the exception of a few allies, look more worrisome than at any point since perhaps the immediate aftermath of World War II.

A Hard Look at Hard Power provides in-depth analysis of the state of key allied militaries. It could not be more timely.

DOUGLAS C. LOVELACE, JR.
Director
Strategic Studies Institute and
    U.S. Army War College Press

CHAPTER 1

INTRODUCTION

Gary J. Schmitt

Since World War II, a key element of America's grand strategy has been its worldwide network of strategic allies and partners. This network has provided the United States with the framework for sustaining its global presence, enhanced deterrence against adversaries in key regions of the world, and, when called upon, provided men and materiel necessary to fight wars. Indeed, since the fall of the Berlin Wall, with one exception — the U.S. invasion of Panama in December 1989 — American forces have not engaged in a major conflict without allies fighting alongside them. Although, in the words of Bill Clinton administration Secretary of State Madeleine Albright, the United States might be "the indispensable nation," as a matter of practice, America is so in conjunction with its security partners.[1]

This practice is grounded in four simple considerations. The first and most straightforward is that allies might have capabilities that increase the overall "punching power" of a given military campaign. Second, allied militaries, even when requiring the assistance of U.S. enablers, will often reduce the overall burden on U.S. forces. Third, and related to the second consideration, is that, when confronted with two major military campaigns as in Iraq and Afghanistan in the last decade, the United States required additional forces to sustain both campaigns simultaneously. As a matter of "economy of force," allied militaries helped "hold" Afghanistan against the Taliban as the body of American military forces turned their attention to

1

the main action in Iraq from 2003 to 2009. And, finally, although U.S. administrations routinely claim the prerogative of acting unilaterally to address threats to U.S. security, the American body politic prefers to act in conjunction with allies—especially democratic allies—when engaging in military operations. It does so for the simple reason that the American public and its leaders believe that coalitions of like-minded liberal governments confers a degree of legitimacy on such operations that unilateral action is short of. Whether this is necessarily the case—and, arguably, unilateral actions can be just as legitimate as those undertaken under "collective security" arrangements in certain circumstances—the political and diplomatic reality is that the United States favors going to war with other democracies.

Despite this preference for coalitions, following the end of the Cold War and the existential threat posed by the Soviet Union and its Iron Curtain allies, increasingly less attention was paid to America's allies—especially their "hard power" capabilities—in the 1990s. Everyone, including the United States, was busy collecting on the "peace dividend" that seemed to flow from the fact that the West was no longer facing a military superpower. To be sure, there were new missions for our European allies, such as in the Balkans and Africa, but those missions did not require militaries of the scale that had previously been under arms. Moreover, savings from cutting the size of the militaries could then be put to modernizing and reshaping them; it would be a "win-win" for America's security partners. Except it was not.

New platforms cost more than expected. Personnel costs for all-volunteer forces continued to rise, and governments continued to expand domestic social programs, squeezing out what little budget space remained for defense spending. Compounding these problems for allies who joined the fight in either Iraq or Afghanistan, or both, was the reality that those campaigns were prolonged, "boots-on-the-ground" intensive, and required equipment and platforms unique to those fights. Toss in economies hard hit by the "great recession" of 2008 and the lackluster recoveries that followed, and one has a recipe for an even further decline in the hard power capabilities of key allies.

North Atlantic Treaty Organization (NATO) allies' continued effort to try to "do more with less" has resulted in a decade-long series of complaints from senior U.S. officials that too many of our allies have not kept to the 2002 agreed-upon benchmark of spending a minimum of 2 percent of their gross domestic product (GDP) on defense. Nor is this a problem confined to NATO and Europe. Key Asian security partners— South Korea, Taiwan, Japan, and Australia—fall below the 2 percent line, as well.

As justified as those complaints are and as useful as it is for generally measuring a country's defense burden, focusing on military spending as a percentage of GDP is insufficient for fully understanding each country's military-strategic plans, capacities, and outlook. The chapters which follow, commissioned over the past few years by the Marilyn Ware Center for Security Studies at the American Enterprise Institute, Washington, DC, are intended to fill in that gap. The chapters, written by country and security experts, examine current and planned defense budgets, troop strengths, deployable capabilities, procurement programs, research and development efforts, doctrinal updates, and strategic guidance documents in an effort to provide an accurate, well-rounded account of various key allies' hard power capabilities.

In addition to the country-specific chapters, there are also chapters that provide an overview of NATO land, air, and maritime forces, and a chapter discussing the possibilities and limitations of the attempt to squeeze more capabilities of allied militaries through "smart defense" and "pooling" initiatives.

This focus on "hard power" is not intended to shortchange the utility of "soft power"—what Harvard professor Joseph Nye has described as being the ability to attract rather than coerce other states into doing what you want. But, as we have seen in Eastern Europe, the Middle East, and East Asia in recent years, the absence of military capabilities or the strategy to deploy them effectively can create regional dynamics that invite instability or, worse, a vacuum that soft power cannot fill by itself.

Having a fuller understanding of allied military capabilities, plans, and strategies is becoming even more important as the U.S. Government cuts its own defense budget and force structure. For American policymakers and strategists, knowing what relative assistance allies and partners can provide now and in the future, will only grow in importance. The chapters that follow are intended to deepen that understanding.

ENDNOTES - CHAPTER 1

1. Madeleine Albright, Interview on NBC-TV, *The Today Show with Matt Lauer*, February 19, 1998, available from *www.state. gov/1997-2001-NOPDFS/statements/1998/980219a.html*.

CHAPTER 2

ITALIAN HARD POWER:
AMBITIONS AND FISCAL REALITIES[1]

Gary J. Schmitt

KEY POINTS

- Although Italy has the eighth largest economy in the world, its military capabilities fall short of key allied countries of similar size and economic strength because of its government's long-term failure to increase its defense budget.
- Facing severe fiscal constraints, the Italian government has issued a new round of defense spending cuts that has substantially lowered overall force structure, but which the government hopes will still allow for continued modernization of its forces.
- The question going forward is whether the regional and global ambitions Rome once had for its military will diminish as its forces contract.

Although recent headlines have highlighted Italy's dire fiscal situation, its defense capabilities have been in decline since well before the latest economic crisis. For Americans who grew up reading about the sometimes poor performance of Italian forces in World War II or watching movies set in Rome in which the theme is *la dolce vita*, perhaps this comes as no surprise.

However, Italy remains one of the world's leading economies; it had the eighth largest gross domestic product (GDP) in 2011.[2] Indeed, in terms of the size of its economy and population, the two nations Italy most resembles are France and the United Kingdom (UK). But, in terms of willingness to turn these attributes into hard military power, Rome falls short of benchmarks set by Paris and London.

As Figure 2-1 elucidates, Italy's defense burden (measured as a percentage of GDP), while never high in the past, has declined even more in recent years.[3] As a percentage of GDP, Italy's defense burden has dropped substantially from what it was just a decade ago—and well below the 2 percent minimum that the North Atlantic Treaty Organization (NATO) allies agreed to try to obtain at the alliance summit in Prague in 2002.

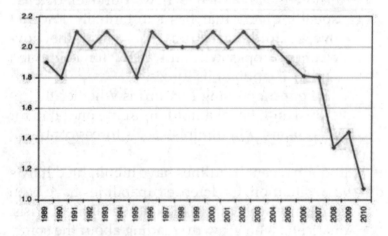

Source: Data derived from "Nota Aggiuntiva allo Stato di Previsione per la Difesa per l'Anno" ("Additional Note to the Defense Budget for the Year"), 2003-2011/2012. Data expressed in current prices.

Figure 2-1. Italy's Defense Expenditure as a Percentage of GDP.

Although both France and the UK have also seen defense spending decline, Italy's per-capita expenditure on defense, according to the Italian defense ministry, lags significantly behind that of its NATO allies (see Figure 2-2). On the face of it, Italy is punching well below its weight (see Figure 2-3).

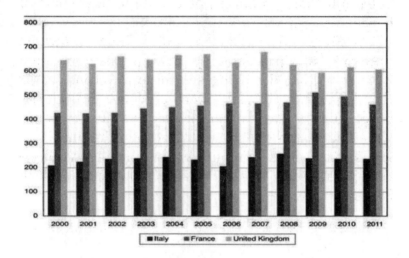

Source: Data Derived from "Nota Aggiuntiva allo Stato di Previsione per la Defesa" (2003-2011/2012). Data expressed in current prices.

Figure 2-2. Defense Spending per Capita (€).

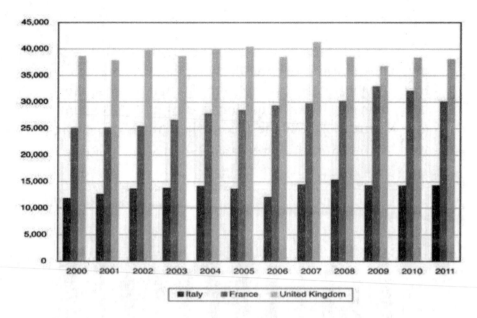

Source: Data derived from International Institute for Strategic Studies, *The Military Balance*, 1991-2012, London, UK: Arundel House.

Figure 2-3. Base Defense Budget (Millions €).

Nor is Italy's defense budget picture improving. According to the Italian defense ministry, its base defense budget [*Funzione Difesa* (FD)]—never large to begin with—will fall to €13.6 billion this year (see Figure 2-4).[4]

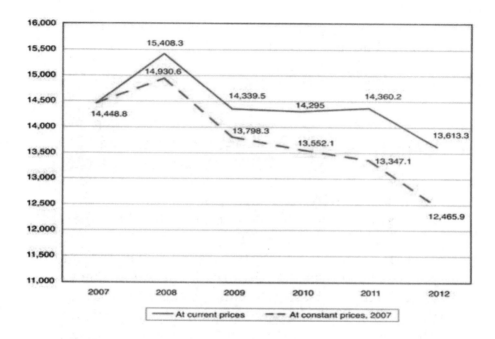

Source: "Nota Aggiuntiva allo Stato di Previsione per la Difesa per l'Anno 2012" ("Additional Note to the Defense Budget for the Year 2012"), p. 140.

Figure 2-4. Italian Base Defense Budget (Millions €).

Compared to the FD average of the previous 4 years (2008–11), this amounts to a cut of some 7 percent. Significantly, the "investment" (procurement) portion of the budget for 2012 has been shorn by 25 percent from the previous 4 years and has seen a drop of nearly 30 percent from 2011 to 2012 alone.

Also important is the reduction in funds allotted to Italy's Ministry of Economic Development, which subsidizes Italian defense research and development and procurement programs, as well as a 30 percent reduction in funds for military operations abroad.

Under new austerity measures, Italy will reduce its defense budget by €3 billion over the next 3 years.[5]

ITALY'S STRATEGIC VISION

Any analysis of Italian grand strategy faces one overriding difficulty: there is no systematic production by the government of national-level strategy papers. To the extent that strategic documents have been issued, more often than not, they have been at the initiative of individual ministers rather than an established policy planning process.

That said, there have been various government papers issued over the past decade that allow one to tease out Italy's strategic ambitions, and the military — the government believes — is required to obtain them. The most relevant documents of this sort have been the 2001 defense ministry's *New Forces for a New Century*; the post–September 11, 2001 (9/11) *Defense White Paper*, issued by the ministry in 2002; the 2005 *Defense Chief of Staff's Strategic Concept* paper; the defense staff's 2005 *Investing in Security: The Armed Forces, An Evolving Tool*; and the defense ministry's annual *Addendum to the Defense Budget*, which attempts to give strategic and political context to the approved budget, as well as provide details on specific accounts within the budget.

The 2001 document was the first formal paper of its kind produced by the Italian defense ministry since the end of the Cold War — indeed, it was the first since the mid–1980s.[6] The paper notes the obvious but important point that Italy will not be facing a conventional military threat to its homeland anytime soon. But it couples that fact with the assertion that Italy's interests are "quite broad," ranging from Southeastern Europe to the Caucasus, from the Horn of Africa to the Maghreb, and that Italy's military contribution to collective security and stabilization efforts in recent years has ranged far, wide, and outside the areas directly affecting Italy's own strategic national interest.[7]

New Forces offers up a relatively ambitious strategic outlook, including Italy potentially having the capability to take the lead in military operations. To meet those ambitions, the paper notes that Italy will need to progress in creating an all-professional military, work with allied countries to develop and produce a plethora of new weapons systems, and increase its defense expenditures from 1.5 to 2.0 percent of GDP. The post–Cold War "peace dividend" had to end if Italy's military was going to be able to handle the expected increased involvement in multilateral (NATO- and European Union [EU]-led) military operations, and do so as a capable allied force.

The 2002 white paper was published in the wake of the 9/11 attacks on the United States and the subsequent removal of the Taliban-led government from power in Afghanistan. Not surprisingly, the paper pays particular attention to the then-emerging threat of Islamist terrorism and, similar to the previous year's document, emphasizes the military's need to operate abroad in concert with allies or under the auspices of the United Nations. With the recent conflict in the Balkans and Afghanistan being on the defense ministry's mind, stabilization missions were at the forefront, leading, among other things, to a potentially enhanced role for the Carabinieri—Italy's national military police force—in peacekeeping operations.

As in 2001, the 2002 white paper reemphasized the need to reform and modernize the Italian military. It noted that the air force was short on modern fighters and even missiles for its planes. It also pointed out that much of the navy fleet was aging, and advocated accelerating the transition to an all-professional Italian military, with a total active duty level for the armed forces set at 190,000, and 12 to 13 smaller but more capable army brigades. To help pay for this transformation, the white paper hoped to find savings in lower overall force structure and a new level of allied defense industrial cooperation to reduce costs while simultaneously increasing interoperability.[8]

The 2005 *Strategic Concept* paper was not a substantial break from previous papers, but it did attempt to provide a somewhat fuller account of the military tasks confronting a European power in the post–Cold War, post–9/11 era. The paper claimed that, in addition to traditional requirements such as protecting the homeland, Italy faces threats that are increasingly "multi-layered and unpredictable," requiring a preemptive military capability and a capacity to intervene rapidly even when the threat is some distance from Italy.[9]

Italian forces will thus need to act more jointly and, more often than not, in concert with allied militaries. To do so, the military will require enhanced command and control capabilities, surveillance assets, mobility, logistic support, and precision-guided weaponry. The Italian military should aim for a qualitative improvement of its capabilities that are more in line with NATO's leading powers and that allow the military to address the wide range of security problems it might be asked to address.

After the *Strategic Concept* paper, the defense staff's *Investing in Security* paper was published.[10] With its focus on the likely requirements for the Italian military over the next 15 years, the document drills down even further than the *Strategic Concept* paper in its matching of specific scenarios with force requirements. It lays out what capabilities it would need to secure "national spaces" and an immediate-reaction expeditionary (land, air, and naval) force that could act as an independent entry force, operate alone for 30 days if necessary, and for 6 months as part of a larger multinational operation.[11]

Also useful for understanding Italy's strategic posture or, more specifically, the connection between the country's ambitions and the military resources it is willing to apply are the yearly *Nota Aggiuntiva allo Stato di Previsione per la Difesa* (Additional Note to the Defense Budget). The "Additional Note" to the defense budget is sent to the Italian parliament under the signature of the defense minister and provides an overview of how the ministry views the overall security situation and, in turn, its plans and programs for the military to meet its security objectives.

Starting with the *Nota Aggiuntiva* for the 2001 budget—a document released in October 2000—and ending with the *Nota Aggiuntiva* for 2012, Italy's post–Cold War view of the security environment has been relatively stable. The notes first and foremost recognize that Italy faces no conventional military threat of any consequence to its homeland. However, since the late-1990s, Italian governments of both the left and right perceive Italy's security as being affected by instability in the Balkans, North Africa, the Horn of Africa, the Middle East, the Mediterranean Basin, and, since the 9/11 attacks, even farther afield.

Hence, the country's security problems are "multidimensional" and of "undefined contours."[12] This, in turn, requires, as note after note suggests, a military that is deployable, flexible, sustainable for extended periods, and modernized so as to be capable of operating in conjunction with top-line forces of both NATO and the EU.

Indeed, at the turn of the century, in the 2001 note, Italy was not shy about its ambitions. With plans to begin reversing the "peace dividend" cuts to the military that took place throughout the 1990s, Minister of Defense Sergio Mattarella declared that Italy's global "credibility" had grown, making Italy "one of the leading" countries in NATO and the EU, as well as, he pointed out, being the fourth largest contributor to UN peacekeeping missions.[13]

With more than 8,000 of its military deployed abroad—ranging from operations in the Balkans to a stabilization mission in East Timor—Italy was asserting itself in a manner that allowed it to increasingly play a role in that group of nations driving international affairs. The 2001 bump in defense spending was only the first step, the note argued, in Italy's military acquiring the kind of capabilities needed to match its ambitions and ensuring that it would not be a "mere spectator" in addressing future security problems.[14]

Indeed, by 2006, more than 10,500 of Italy's military were deployed abroad, including to Iraq and Afghanistan. While the numbers were "unprecedented" for post–World War II Italy, the note also stipulated that, in light of the generally unpredictable security environment, those numbers could no longer be thought of as "unusual."[15]

Of course, increasing deployments abroad, while at the same time modernizing Italian forces, required greater resources for Italy to fulfill its new strategic ambitions. As with most European states following the end of the Cold War, Italy had made deep cuts in its defense budget. The increase in defense spending in 2001 was meant to be the first step in reversing course and, eventually, putting Italy on par with France and the UK when it came to defense spending and military credibility.

According to the note attached to the 2002 budget, the goal was to have the base defense budget (FD) equal 1.5 percent of GDP and then be sustained there.[16] At that level, the FD would be more or less aligned with other "major" European allies. However, this would require a change in Italian spending priorities since, in 2002, the FD was less than 1.1 percent of the country's GDP.

Even with the slight bump in resources in 2002, however, the increase in personnel costs was squeezing the training, maintenance, and investment accounts. Indeed, by 2006, more than 70 percent of the base defense budget was going to personnel costs — far from the "model" allocation in which 50 percent goes to personnel costs, 25 percent is spent on maintaining the force, and 25 percent is spent on procurement and recapitalization.

Further complicating matters was the fact that, between 2002 and 2006, the defense budget was cut every year. By 2006, the base defense budget was down to 0.82 percent of GDP, and the ministry began announcing delays in modernization plans and increasing problems in sustaining the overall readiness of the force. After an increase in the defense budget in 2007 — but, according to the note, just enough of one to support the most pressing operational requirements for overseas operations and to "only partially allow" the ministry to deal with "the already difficult" problem of too few resources — the tsunami of the global economic crisis hit.[17] By 2009, the note was warning that, if the downward direction of the budget continued, the ministry would have to slash the size of its force by tens of thousands, plans for modernization would dramatically slow, and "important programs" would need to be "reduced or postponed."[18] The trend has not been reversed, and, as predicted in the 2012 note, the ministry has formalized plans to shrink the Italian military by 40,000 and cut back or delay procurement programs designed to modernize Italy's military.

In short, since 2007, resources for training and modernization have dropped by over 40 percent and 30 percent, respectively. Like other European states that are reducing numbers of people and platforms, the pledge is that Italy's military will be "of smaller dimensions but with higher quality."[19] Whether that will happen remains to be seen.

But the ambitions Italy set for itself a little more than a decade ago cannot, as the ministry itself made clear from the start, be fulfilled in the absence of a sustained increase in defense funds. In this context, was the fact that Italy was forced to withdraw its aircraft carrier — the *Garibaldi* — from ongoing NATO operations against Libya in July 2011 in order to cut costs the low point from which the Italian forces will now move forward, or a harbinger of things to come?[20]

ITALY'S MILITARY ABROAD

Italy's military during the Cold War was principally focused on defending the country itself. This strategic posture was reinforced by the fact that, as one of World War II's defeated Axis powers, Italy was reluctant (like post-war Japan and Germany) to be viewed as believing that its military was for anything but defending the homeland proper.

To a very limited degree, this attitude toward the use of the military has changed in Japan in the wake of 9/11. Judging by Berlin's use of the military in Kosovo, Afghanistan, and the Horn of Africa over the past decade and a half, it appears that Germany has modified its views about what constitutes a legitimate use of military force. So too Italy, if judging by the number of times its military has been involved in operations outside its borders.[21] Italian forces were sent to Iraq during the first Gulf War, followed shortly thereafter by a deployment to Somalia, and then to Bosnia. Other deployments have included operations in Central Africa, East Timor, Mozambique, the Balkans, Iraq again, Afghanistan, Lebanon, and, most recently, against Libya, where the Italian Air Force flew air defense suppression and strike missions and helped enforce the United Nations (UN)-sanctioned no-fly zone over the country.

While the activity level of the Italian military has certainly picked up in recent years, perhaps the origins of this new attitude toward using the military dates to 1982 when Italy—along with France and the United States—sent troops into Lebanon in the wake of the First Lebanon War between Israel and the Palestine Liberation Organization and Syria. The deployment arose because Rome believed that, given its geographic location, Italy should have a more prominent role in Middle Eastern and Mediterranean security affairs.

But it was not until Italy's participation in Operation DESERT STORM in 1991—the first time the Italian Air Force had been involved in actual military operations since World War II—that the rate of the military's deployments surged and appeared to open the door to more kinetic use of force. For example, in the 1991 Kosovo War air campaign against Yugoslav forces, Italy was the third largest contributor of aircraft and flew the fourth largest number of sorties by a NATO member.[22]

However, more recent deployments present a mixed picture when it comes to the use of military force, especially in the cases of Italian ground contingents sent to Iraq in 2003 after Saddam Hussein was removed from power, to Afghanistan as part of the International Security Assistance Force (ISAF) mission, and to Lebanon following the "33-Day War" between Israel and Hezbollah in 2006. Wary of casualties and unwilling to provide the extended security rationale that would be needed to justify Italy's involvement in all three missions, successive governments in Rome have sold these deployments—involving thousands of Italian soldiers in total—to the Italian public as "peacekeeping" and "humanitarian" missions. But, of course, neither the Iraq nor the Afghanistan mission turned out to be the "soft" power, light security missions the Italians expected.

Iraq.

The Italian military's deployment to Iraq—which lasted from June 2003 until November 2006—was certainly as difficult an experience for Italy's forces as what they faced in Afghanistan, and undoubtedly reinforced Rome's inclination to take a cautious operational approach in Afghanistan. Coming on the heels of the American-led military campaign removing Saddam Hussein from power—a campaign decidedly unpopular with the Italian electorate—the decision to send Italian troops was justified by the government as an "urgent intervention in favor of the Iraqi people." Keeping with this theme, Italy's defense minister at the time said the intervention was just the "opposite of war."[23]

But war it was. Just a few short months after deploying almost 3,000 troops to Nasiriyah, a city in Dhi Qar Province southeast of Baghdad, a lightly protected Italian outpost was attacked by a suicide bomber, which killed 13 Italian military policemen and four soldiers. In response, the order was given to move most of Italy's forces out of the city. This was not going to be the kind of "peacekeeping" and "stabilization" operation Italian forces had previously conducted in the Balkans.

Indeed, throughout the spring of 2004, Italian forces were engaged in a form of urban warfare with the Mahdi Army, as this Shia militia attempted to take advantage of Rome's decision to reduce its footprint in the city.[24] Lacking firepower, numbers, sufficiently armored vehicles, and surveillance capabilities, the best the Italian forces could do was establish a strategic standoff for control of the city. Eventually, the decision was made to concentrate the vast bulk of Italian troops at Tallil Air Base outside the city. With the change of government in Rome in April 2006, the decision was made to end the Iraq mission altogether.

Afghanistan.

There is little question that the Italian military's involvement in Afghanistan has been the largest, most complex, and most difficult campaign for the country since World War II. A little over 2 months after the 9/11 attacks, elements of Italy's navy (an aircraft carrier, two frigates, and a tanker) were steaming toward the Indian Ocean in support of Operation ENDURING FREEDOM (OEF). Engaged principally in sea-control duties and at-sea inspections of suspicious vessels, the carrier *Garibaldi* deployed with eight AV-8 (Harrier) ground-attack jets that flew nearly 300 missions over Afghanistan. However, Rome had restricted the Harriers' use to target identification, leaving actual strike missions to other allied planes.[25]

On the ground, Italy's contribution to ISAF has recently topped 4,000 troops (see Figure 2-5). In addition, the Italian military assumed overall command of ISAF from September 2005 to May 2006, took the ISAF lead in 2005 of the geographically large and forbidding area of western Afghanistan, headed up the Provisional Reconstruction Team in Herat, and contributed forces to several mentoring teams tasked with training Afghan security forces by partnering with them in the field.

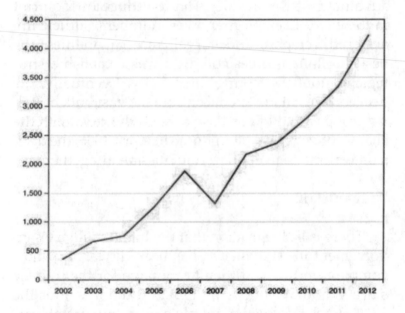

Source: Data derived from *The Military Balance*, 2002-12, International Institute for Strategic Studies, London, UK: Arundel House.

Figure 2-5. Deployment of Italian Troops in Afghanistan.

While all this information is well known, very little has been written about combat operations involving Italian forces. At least initially, this was due to the fact that the troops sent to Afghanistan were lightly armed and equipped as though their mission would be Kosovo-like peacekeeping. Indeed, the original UN-sanctioned ISAF mission, as opposed to the OEF effort to overthrow the Taliban and hunt down al-Qaeda remnants, was understood as having the more limited mandate of providing security to support efforts at re-building the Afghan state. The last thing Rome wanted to talk about was the idea that "providing security" might require more robust military operations.

This is not to say that Italian troops have not been involved in kinetic operations. For example, from mid–March 2003 to mid–September 2003, a contingent of 1,000 Italian troops was involved in Operation NIB-BIO. Operating out of a base in Paktia, a province on the border with Afghanistan, the Italian forces were tasked with helping coalition forces disrupt efforts by al-Qaeda and the Taliban to reinsert themselves into this heavily Pashtun area.[26]

However, the Italian government having sent them — without helicopters, heavy weaponry, or ar-mored land transport — to eastern Afghanistan, there was a limit to what Italian forces could do. As a result, the bulk of their efforts consisted of setting up check-points, establishing blocking positions at potential insurgent escape routes, and conducting intelligence-gathering patrols.

Although RC-West (the ISAF designation for the four provinces of Herat, Farah, Badghis, and Ghor over which Italy's military had overall command for the region) was not a hotbed of Taliban activity by 2006, insurgent activity was increasing in the region.

But in an area nearly half the size of Italy and containing more than 2.5 million Afghans, the resources the Italian forces had been provided in manpower, firepower, and transport meant that, even in conjunction with allied forces in the region, fulfilling the ISAF mission of "securing" the region became an increasingly improbable task.

As a result of pressure from both its own military and ISAF allies, Rome did increase the size of the Italian force in RC-West and provided more assistance in terms of armor, jet aircraft, air transport, unmanned aerial vehicles, and attack helicopters. This gave the Italians a greater capacity to engage in blocking operations as the Taliban fled from ISAF operations in nearby Helmand into RC-West and, in a limited number of cases, to participate in operations designed to clear pockets of Taliban in their area of responsibility.

Nevertheless, it is also the case that Italian governments — both of the left and the right — have not wanted Italian soldiers to participate in operations in the more dangerous areas in the south or the east regions of the country. It was only in 2008 that the Italian government modified its caveat that Rome would have to approve any and all requests for Italian forces to assist coalition forces outside of RC-West by lowering the time allotted for it to respond from 72 hours to 6.

As with other ISAF contributors, Italy has begun to draw down the numbers deployed to Afghanistan. Because it is pressed financially, Rome would like to reduce the Italian deployment by 1,200 over the next year and gradually wind down force levels to no more than 800 to 1000 troops in country by the end of 2014, with 2014 being the year the Afghan government takes the lead in providing security throughout the country.[27]

Although the Italian military's experience in Iraq and Afghanistan can, at best, be described as problematic, there is no question that the deployments have helped the Italian military in its goal of creating a more professional force. Working with allies in a hostile environment far from Italy has forced Italian forces to "up their game" when it comes to training, logistics, and field-level modernization.

Whether the same can be said for the Italian policymakers who decide how to employ Italy's military abroad and provide the rationale for doing so is a different question. As former chief of Italy's defense staff General Mario Arpino pointedly remarked regarding the mission in Afghanistan: "If Italy participates in international missions just to be there, to get a little prestige, but without understanding what the dangers are . . . we risk doing damage to the interests of our country."[28]

MOVING FORWARD

As noted previously, in 2012, the Italian government proposed plans to restructure its defense effort to keep it more in line with the resources at hand. According to the defense ministry's note for this year, "Today's reality [is marked by a] significant imbalance" between personnel costs and the monies available to keep the military trained, ready, and modernized.[29]

The heart of the plan is to reduce personnel costs, now more than 70 percent of the base defense budget (see Figure 2-6), by dropping the active duty numbers authorized from 190,000 to 150,000, and by slicing the civilian work force to 20,000 from its current 30,000. With the cut in force structure, expected savings from eliminated military overhead and the sale of no-longer-needed infrastructure, the hope is to free up resources for the "investment" and "training" accounts. If successful, the budget's general parameters would be more in line with what the defense ministry calls its "most significant allies," meaning 50 percent would go to personnel, 25 percent to modernization, and 25 percent to training and maintenance.

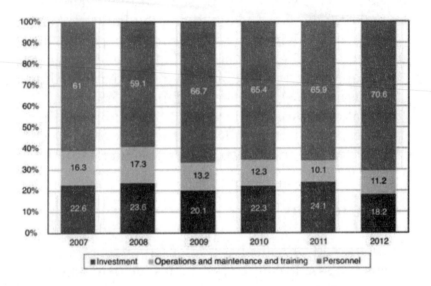

Source: "Nota Aggiuntiva allo Stato di Previsione per la Defesa per l'Anno, 2012" ("Additional Note to the Defense Budget for the Year 2012"), p. 142.

Figure 2-6. Breakdown of the Defense Function by Spending Area.

In the short term, however, the defense investment account is taking a beating, with a reduction in spending of 28 percent from 2011 to 2012. (For the individual services and their respective investment budgets, this means a cut of 52 percent for the army, 40 percent for the navy, and 29 percent for the air force.) Given the 30 percent reduction in defense investments since 2007, the modernization hole is deep and will require a substantial effort to be dug out of.

Moreover, while the 2012 budget increased spending on operations and maintenance and training by 5.4 percent, since 2007, spending in this area had fallen by 40 percent—again, a deep hole to climb out of. According to commander of the Italian joint operations headquarters General Marco Bertolini, if funds for training were not boosted, Italy would not be able to undertake another mission like Afghanistan; or, as the defense ministry itself notes more prosaically, this year's increase will still be "insufficient" to meet the services' needs.[30]

As for the Italian Air Force, the budget reductions have substantially reduced the number of fourth and fifth generation fighter aircraft it will be flying. A decade ago, the initial goal was to replace Italy's aging fleet of F-104s, AMX fighter bombers, and leased F-16s with a buy of 121 Eurofighter Typhoons, 40 F-35Bs, and 69 F-35As. The Typhoon order has, however, been cut back to 96, with some 62 now in service; the F-35B buy reduced to 15; and the F-35A purchase pared back by 9.[31] Although these new acquisitions will clearly be an upgrade in individual aircraft capabilities, the fleet itself has declined from 313 fighter aircraft in 2001 to 220 today and, once the 70 or so multirole, 1970s-designed Tornadoes are retired from service over the next decade, the Italian tactical fighter fleet could consist of only 150 aircraft.

The Italian Navy is following a similar path. In June 2012, navy chief Admiral Luigi Binelli Mantelli announced that 26 or 28 vessels would be retired over the next half-decade. Although new and more capable platforms will be added to the fleet, overall numbers will drop as the replacement vessels will not be 1:1 for those withdrawn from service. Indeed, to save the cost of decommissioning the ships, the government is looking to sell them at a discount to other countries or, even, to simply give them away.[32] Examples of the cuts include reducing the submarine force from the current six to four (about half the number in 2001), dropping the number of new frigates to be bought from 10 to 6 (leaving the total number of frigates at 10 after seven or eight older frigates are pulled from service), cutting minesweepers from 12 to 8, and patrol boats from 18 to 10. Moreover, plans for replacing the retiring carrier *Garibaldi* and amphibious transport docks with the much larger carrier *Cavour* and amphibious assault ships (LHDs) has been complicated by a reduced buy of F-35Bs and the freezing of the LHDs' acquisition.[33]

The number of army combat brigades has also shrunk. In 1991, there were 19 combat brigades. By 1997, the number had dropped to 13. Under the new plan, the combat brigades will go from the current 11 to 9. Concurrently, the Italian army has seen the number of tanks cut by more than half since 2001, with an equally substantial loss in numbers of field artillery and mortars. Smaller and less "heavy," the army hopes to use the savings from fielding a leaner force to upgrade its fleet of attack helicopters, increase the capabilities of its special operations forces, and modernize its inventory of land vehicles.

To maximize the effectiveness of its smaller armed forces, the ministry's plan is to invest in greater service jointness; enhanced command, control, communications, computers, and intelligence (C4I) capabilities; a digitalized (net-centric) land force; and upgraded surveillance and target acquisition systems for the navy and air force. And, indeed, other than line items for expenditures on the F-35 program (€548.7 billion) and the final two U-212 submarines (€170.7 billion), the two most expensive programs listed in the defense budget are for programs involving C4I and ground surveillance (€160 billion) and jointness (€154 billion).[33]

Even so, the efforts to enhance the effectiveness of the smaller force despite budget cuts have meant some important programs have "slipped." For example, delivery of the last pair of U-212 submarines has been pushed back a year, while the time frame for the planned procurement of medium armored vehicles, multirole helicopters, and various advanced munitions has been shuffled to the right by 2 to 4 years. This collectively suggests that, even with the substantial cuts in the overall size of the military, the number of civilian employees, and no-longer-needed military infrastructure, the margin of error for Minister of Defense Di Paola's vision of creating a smaller but better equipped and advanced military is a thin one. Unexpected cost increases for major programs, fewer savings from personnel and infrastructure reductions, or further cuts in defense spending to address current deficits in government spending could undercut his plans for Italy's military.

CONCLUSIONS

Under current plans, Italy's military will retain a wide spectrum of capabilities befitting a medium-sized global power. As such, according to Di Paola, the government will not only have sufficient "hard power" to ensure Italy's own defense, but a range of military tools from which Rome can pick and choose how it will involve the country in operations abroad. But without strategic airlift and sealift, Italy will, in most instances, either require a relatively permissive environment to deploy a substantial number of forces or the assistance of NATO allies. Moreover, with cuts in numbers to personnel, platforms, and resources, Italian policymakers will find they have less discretion in where and when they use the military. While the forces themselves might be more capable, a smaller military in a tight fiscal environment will inevitably lead Rome to conserve the capabilities it has.

A decade ago, Rome acted on the unstated but implied quid pro quo that, in exchange for U.S. and allied assistance in stabilizing the Balkans, ensuring energy supplies from the Persian Gulf, and keeping Islamist terrorism at bay, Italy would offer military assistance in places like Iraq and Afghanistan—locations which the Italian public, however, did not readily consider vital to Italy's national security. This dynamic also fit Rome's sense that it could and should play a larger role on the world stage. But that implicit deal and the ambition that accompanied it have gradually come undone in the face of fiscal pressures and the public sense that Italy did not face the kind of immediate threats that required maintaining, let alone increasing, Italy's defense burden.

However, if the United States follows through on its decision to focus more of its attention on ensuring a favorable military balance in the Asia-Pacific region, and does so by reducing its military footprint in Europe, then countries such as Italy will be expected to do more in meeting their own security needs. Those security tasks appear to be growing, not receding. Not only is Iran's threat to stability in the Gulf increasing, but the Horn of Africa and large segments of the

Mediterranean Basin appear less and less stable—all of which could, and probably will, impact Italy's security. But with military spending cut to the bone, Italy's ability to help address those challenges will likely fall short not only of what one might expect of a country its size and economic weight, but also of Rome's own ambitions at the century's turn.

ENDNOTES - CHAPTER 2

1. This essay was originally published on November 1, 2012.

2. World Bank, "Gross Domestic Product, 2011," available from *databank.worldbank.org/databank/download/GDP.pdf*.

3. Determining Italy's level of defense spending is complicated by the fact that the government's overall defense budget includes funds for Italy's national military police, the Carabinieri—a force of some 100,000. Approximately a quarter of Italy's overall defense budget goes to pay for the Carabinieri and the internal security function. At the same time, the defense budget does not include funds spent on overseas operations, such as in Afghanistan; those funds are approved separately by Italy's parliament. Nor does the nominal defense budget include funds spent by Italy's Ministry of Economic Development on military procurement and research and development. Although the ministry does not publish the exact amount it contributes to defense-related spending, estimates were that, in 2011, the ministry spent €1.85 billion, and, in 2012, the total will drop to somewhere between €1.30 and €1.67 billion.

4. The FD includes funding for the three military services (army, navy, and air force), training and maintenance, personnel, and weapons development and procurement.

5. Andy Nativi Genoa, "Italy Wants More For Less With Defense Cuts," *Aviation Week and Space Technology*, September 3, 2012, available from *www.aviationweek.com/Article.aspx?id=/article-xml/AW_09_03_2012_p16-486416.xml&p=1*.

6. The last white paper was published in 1985. See Maurizio Cremasco, "Italy: A New Role in the Mediterranean?" John Chipman, ed., *NATO's Southern Allies*, London, UK: Routledge, 1988.

7. Italian Ministry of Defense, *Nuove Forze per un Nuovo Secolo* (*A New Force for a New Century*), Rome, Italy, 2001, available from *www.difesa.it/InformazioniDellaDifesa/periodico/IlPeriodico_Anni-Precedenti/Documents/2001_-_Nuove_forze_per_un_nuovo_secolo. pdf#search=Nuove%20Forze%20.*

8. Italian Ministry of Defense, *Libro Bianco* (*2002 White Paper*), 2002, Rome, Italy, December 20, 2001, available from *www.difesa. it/Approfondimenti/ArchivioApprofondimenti/Libro_Bianco/Pagine/ Premessa.aspx.*

9. Italian Ministry of Defense, *Il Concetto Stategico del Capo di Stato Maggiore della Difesa* (*The Chief of the Italian Defence Staff Strategic Concept*), Piedimonte Matese, Italy, April 2005, available from *www.aeronautica.difesa.it/Missione/Documents/libroconcetto-strategico.pdf.*

10. Italian Ministry of Defense, *Investire in Sicurezza: Forze Armate: Uno Strumento in Evoluzione* (*Invest in Security: The Armed Forces: An Evolving Tool*), Piedimonte Matese, Italy, October 2005, available from *www.difesa.it/InformazioniDellaDifesa/ periodico/IlPeriodico_AnniPrecedenti/Documents/Investire_in_ sicurezza.pdf#search=Investire%20in%20Sicurezza.*

11. For the army, this meant an Italian combat land force consisting of 10 brigades (three light, three medium, three heavy, and one airborne), with planned modernization of ground combat vehicles and attack and transport helicopters. For the navy, it meant a smaller but more capable fleet, with larger amphibious transport ships, a new aircraft carrier flying F-35Bs, replacing an existing carrier with new amphibious assault ships, and new diesel submarines. For the air force, the future fleet would eventually consist of a mix of Eurofighter Typhoons and F-35s, and a sustained deployment capability of 45-50 Aircraft, with four new KC-767 aerial refueling tankers. See Italian Ministry of Defense, *Ivestire in Sicurezza.*

12. For characterization of Italy's threat environment as "multidimensional," see *Il Concetto Stategico*, p. 10.

13. Italian Ministry of Defense, *Nota Aggiuntiva allo Stato di Previsione per la Difesa per l'Anno 2001* (*Additional Note to the Defense Budget for the Year 2001*), October 2001, available from *www.difesa.it/Approfondimenti/Nota-aggiuntiva/Documents/58570_na_2001.pdf*.

14. *Ibid.*

15. Italian Ministry of Defense, *Nota Aggiuntiva allo Stato di Previsione per la Difesa per l'Anno 2006*, 2006, available from *www.difesa.it/Approfondimenti/Nota-aggiuntiva/Documents/86808_Nota-Aggiuntiva2006.pdf*.

16. Italian Ministry of Defense, *Nota Aggiuntiva allo Stato di Previsione per la Difesa per l'Anno 2002* (presented to the Italian Senate March 18, 2002), available from *www.difesa.it/Approfondimenti/Nota-aggiuntiva/Documents/22689_notaaggiuntiva2002.pdf*.

17. Italian Ministry of Defense, *Nota Aggiuntiva allo Stato di Previsione per la Difesa per l'Anno 2007* (presented to the Italian Senate May 15, 2007), available from *www.difesa.it/Content/Documents/nota_aggiuntiva/86808_NotaAggiuntiva2007.pdf*, accessed on September 25, 2012.

18. Italian Ministry of Defense, *Nota Aggiuntiva allo Stato di Previsione per la Difesa per l'Anno 2009* , 2009, available from *www.difesa.it/Content/Documents/nota_aggiuntiva/3171_Nota_Aggiuntiva_2009.pdf*, accessed on September 25, 2012.

19. Italian Ministry of Defense, *Nota Aggiuntiva allo Stato di Previsione per la Difesa per l'Anno 2012* (presented to the Italian Senate in April 2012), available from *www.difesa.it/Approfondimenti/Nota-aggiuntiva/Documents/NotaAggiuntiva2012.pdf*, accessed on September 1, 2012.

20. "Italy Removes Aircraft Carrier from Libya Campaign," *Defense News*, July 7, 2011, available from *www.defensenews.com/article/20110707/DEFSECT05/107070311/Italy-Removes-Aircraft-Carrier-from-Libya-Campaign*.

21. For an overview of Italy's military operations abroad, see Piero Ignazi, Giampiero Giacomello, and Fabrizio Coticchia, *Italian Military Operations Abroad; Just Don't Call it War*, New York: Palgrave Macmillan, 2012.

22. John E. Peters *et al.*, *European Contributions to Operation Allied Force: Implications for Transatlantic Cooperation*, Santa Monica, CA: RAND Corporation, 2001, p. 21, available from *www.rand.org/ content/dam/rand/pubs/monograph_reports/2007/MR1391.pdf*.

23. Ignazi, Giacomello, and Coticchia, pp. 76, 140–141. It is useful to remember in this context that before an international mission is undertaken by Italian forces, the parliament votes on the government's proposal, which includes the language for the mission's mandate and the size of the force to be sent.

24. Riccardo Cappelli, "Iraq: Italian Lessons Learned," *Military Review*, March–April 2005, available from *www.scribd. com/doc/8352865/Iraq-Italian-Lessons-Learned-Military-Review-with-errata*.

25. Gianandrea Gaiani, *Iraq-Afghanistan: Guerre di Pace Italiane* (*Iraq-Afghanistan: Wars of Italian Peace*), Venice, Italy: Studio LT2, 2008.

26. Task Force Nibbio, General Office of the Chief of Staff Office of Public Information, Italian Ministry of Defense, October 13, 2003, available from *www.difesa.it/OperazioniMilitari/op_int_ concluse/Afghanistan_Nibbio/Documents/92952_SchedaNIB-BIO131003.pdf*, accessed on September 7, 2012.

27. "Italian Defence Minister Visits Afghan Bases as Troops Prepare Handover," BBC Monitoring Service Europe, September 25, 2012.

28. Vincenzo Nigro, "*Arpino: Afghanistan, le Critiche si Fanno a Porte Chiuse*" (*Arpino: The Criticisms of Afghanistan Take Place behind Closed Doors*), *La Repubblica*, July 27, 2007, available from *ricerca.repubblica.it/repubblica/archivio/repubblica/2007/07/27/arpino-afghanistan-le-critiche-si-fanno-porte.html*.

29. *Nota Aggiuntiva allo Stato di Previsione per la Difesa per l'Anno,* 2012.

30. Tom Kington, "Finmeccanica CEO Warns of Italian Defense Cuts' Consequences," *Defense News,* September 28, 2012, available from *www.defensenews.com/article/20120927/DE-FREG01/309270005/Finmeccanica-CEO-Warns-Italian-Defense-Cuts-8217-Consequences.*

31. Tom Kington, "Italian AF, Navy Head for F-35B Showdown," *Defense News,* May 15, 2012, available from *www.defensenews.com/article/20120515/DEFREG01/305150010/Italian-AF-Navy-Head-F-35B-Showdown.*

32. Tom Kington, "Italy Looks to Sell Used Ships, Vehicles," *Defense News,* June 4, 2012.

33. The minimum number of F-35Bs required to outfit the carrier was 22, according to the navy. In an agreement struck with the air force, the navy will be able to borrow F-35Bs from the air force inventory, if necessary, during a deployment. See Tom Kington and Vago Muradian, "F-35 Base-Sharing Plan Defuses Spat between Italy's AF, Navy," *Defense News,* July 2, 2012, available from *www.defensenews.com/article/20120702/DEFREG01/307020002/F-35-Base-Sharing-Plan-Defuses-Spat-Between-Italy-8217-s-AF-Navy.*

34. *Nota Aggiuntiva allo Stato di Previsione per la Difesa per l'Anno,* pp. II-1; C/2–14.

CHAPTER 3

AUSTRALIAN DEFENSE IN THE ERA OF AUSTERITY: MIND THE EXPECTATION GAP[1]

Andrew Shearer

The views expressed in this chapter are the author's and in no way reflect the position of the State Government of Victoria.

KEY POINTS

- After sustaining a 10 percent cut in 2012, Australia's defense budget is unlikely to exceed 1.7 percent of gross domestic product (GDP) over the next 5 years.
- As China and other regional powers procure advanced weapons systems, Australia risks losing its long-standing capability edge in key categories such as naval warfare and air combat, a risk that will only be exacerbated by future shortfalls in planned spending.
- Australia must make a commitment to boosting its military capabilities to ensure that it can make a credible contribution in the unilateral and multilateral defense roles it has signed up for.

Like many Western countries, Australia looked for a peace dividend when the Cold War ended. Defense spending fell, ground forces in particular were cut, and key capabilities such as strategic lift were allowed to wither. By the mid-1990s, Australia had only four undergunned and understrength infantry battalions.

It made a token contribution (two frigates, a supply ship, and a handful of medical and other support personnel) to the first Gulf War in 1991 and provided peacekeeping forces, most notably in Cambodia. But the military participated in no major combat operations for more than 2 decades after the Vietnam War. High-end defense capabilities such as antisubmarine warfare were starved of funds and training opportunities and were allowed to atrophy.

Before coming to office in 1996, former prime minister John Howard had been a strong critic of the preceding government's underfunding of defense. At the start of the Howard government's tenure, it made extensive spending cuts to restore the national budget to a surplus, but deliberately quarantined defense. The real watershed for defense, however, came with Australia's leadership of the International Force for East Timor, the regional coalition that in 1999 intervened to restore order in East Timor.

The East Timor operation was a major military, diplomatic, and political risk for an Australian government that was relatively inexperienced in international affairs. Notwithstanding a United Nations mandate for the operation, opposition by rogue Indonesian military units or even inadvertent conflict with Indonesia could not be ruled out. These outcomes were avoided, and the operation was judged a success. But the Australian government was alarmed by the capability gaps revealed by the operation—in particular, the shortcomings in what was needed to deploy and sustain a modest expeditionary force even a short distance from Australia.

The result was the 2000 Australian defense white paper that committed to grow the defense budget by an average of 3 percent per year, in real terms, over the following decade.² It outlined a 10-year plan to boost air, maritime, and strike capabilities and to ensure that Australia could sustain a brigade-sized force on operations for an extended period, while still having a smaller reserve available for other contingencies. During the Howard government's years in office, Australian defense spending increased by 47 percent in real terms and approached 2 percent as a proportion of GDP.

The 2000 white paper highlighted the importance of the United States to Asia-Pacific security while also flagging that China was likely to pose challenges for the U.S. strategic role in the future. It likewise emphasized that, through the alliance, Australia gained invaluable access to U.S. military technology, intelligence, and training opportunities. As a result, the Howard government placed a premium on interoperability with the United States when the government made major defense acquisition decisions. However, the September 11, 2001 (9/11) attacks and the Australian government's response—which included invoking the Australia, New Zealand, United States Security Treaty (ANZUS) for the first time and committing air, naval, and special forces to coalition military operations against al-Qaeda and the Taliban—took Australia-U.S. military and intelligence cooperation to a new level.

Australian Defence Force (ADF) participation in U.S.-led coalition operations in the global war on terror saw Australian air, naval, and special forces operate more closely with their U.S. counterparts than at any time since Vietnam and across a much larger and vastly more complex area of operations. The sharing of intelligence and access to battlefield information systems between the two countries reached unprecedented levels.[3] There were, however, limits to the ADF's contribution. Australia lacked the full range of capabilities, particularly those enablers necessary to deploy and sustain conventional ground forces at (or above) battalion strength in major combat or stabilization operations in Afghanistan or Iraq.

In 2007, Kevin Rudd's government sought to differentiate itself from Howard's by opposing Australia's military involvement in Iraq; however, it offset this by sustaining Australia's troop contribution in Afghanistan and by reaffirming its strong support for the U.S. alliance. The Rudd government's 2009 defense white paper extended Howard's 3 percent real growth spending target to 2017–18.[4] It also called for a "more potent and heavier" maritime force by 2030, including a fleet of 12 larger and more capable submarines. It also emphasized the need for the ADF to strengthen its offensive strike capabilities; modernize its intelligence, surveillance, and reconnaissance (ISR) systems; and expand its cyber warfare capacity.

In the white paper and the accompanying media briefing, the government clarified that the major driver of these decisions was the regional uncertainty caused by China's rapid military modernization. It went further than previous Australian governments in publicly querying the strategic intentions underlying Beijing's rapid acquisition of blue-water naval capabilities and in calling for greater transparency regarding China's defense plans.[5]

However, the 2009 white paper was undermined from the outset by a mismatch between strategic aspirations and capacity to pay for them. The document provided a credible analysis of the regional security environment and a force structure to match, but the funding commitments were weakly rooted.[6] The document's strong association with Rudd became a further vulnerability when the Labor Party peremptorily replaced him as prime minister with Julia Gillard in June 2010.

The extent of the 2009 white paper's overreach became obvious in the 2012 budget, when the politically and fiscally embattled minority Gillard government slashed defense spending by 10 percent—the largest reduction since the end of the Korean War. This followed a 5 percent cut the year before.[7] A total of Australian dollars (AUD)5.5 billion was stripped from the budget over 4 years, including AUD3 billion in reductions for new military equipment and AUD1.2 billion in facilities construction. Equipment procurement was further reduced by AUD2.9 billion as a result of government reallocations. Consequently, Australia's defense spending fell to 1.56 percent of GDP—the lowest level since 1938. Faced with this obvious gap in strategic vision and available resources, the Gillard government moved up the scheduled 5-year defense review from 2014 to 2013.

2013: PAPERING OVER THE CHASM

The defense planners who drafted the 2013 white paper faced the unenviable task of repairing the view that the government was not serious about the country's defenses. However, the paper's proximity to the forthcoming Australian election on September 7, 2013, means it has inevitably been interpreted as a political as much as a strategic document.

The 2013 white paper's greatest distinguishing factor is its tone regarding China's growing regional influence. In contrast to its 2009 predecessor, the newest paper proclaims that "Australia welcomes China's rise" and, rather defensively, that "the Government does not approach China as an adversary."[8] While acknowledging that "China's defence budget continues to record significant year-on-year increases," it describes China's ensuing rapid military modernization as "a natural and legitimate outcome of its economic growth."[9] It highlights the China–U.S. relationship as the single most important determinant of Australia's strategic environment in coming decades and forecasts that a degree of Sino-American competition is inevitable.

But the 2013 paper concludes (without much compelling evidence) that "Australia sees the most likely future as one in which the United States and China are able to maintain a constructive relationship encompassing both competition and cooperation."[10] It also emphasizes Australia's commitment to pursue "strong and positive" defense relations with China, including annual defense talks, ministerial-level strategic discussions, working-level exchanges, and humanitarian and disaster-relief exercises.[11]

The message was not lost on Beijing: a Chinese foreign ministry spokeswoman said the white paper shows "respect" for Australia's relationship with China and expressed hope that it marked a "turning point" in Australian attitudes.[12] China's continued maritime assertiveness in the South China Sea and in waters disputed with Japan, as well as the Australian public's deep-seated ambivalence about aspects of China's rise, mean that this is unlikely.[13] But the fact that Australia has toned down its official public position on China's military modernization represents a significant tactical victory for Beijing in the Western Pacific.

The second prominent theme of the 2013 white paper is its *fin de siècle* emphasis on the drawdown of long-standing ADF contributions in East Timor (withdrawn in March 2013), the Solomon Islands (withdrawn in mid-2013), and Afghanistan (withdrawn by the end of 2013). Former prime minister Gillard emphasized this drawdown when she declared the end of the 9/11 era.[14]

The white paper anticipates that the drawdowns will allow the ADF to refocus its efforts on stabilization and humanitarian assistance operations in Australia's immediate region and on enhancing the ADF's presence in northern and northwestern Australia, where much of Australia's natural resources wealth is located.[15] This echoes U.S. President Barack Obama's Middle East drawdown and "pivot" to Asia, with perhaps similar wishful thinking that Australia's national interests can be circumscribed to its immediate neighborhood and that tomorrow's threats to Australia's security can be divined today.

The third change in emphasis in the 2013 white paper was the adoption of the Indo-Pacific as an organizing principle for Australian strategic policy. It confirms that "The Indian Ocean will increasingly feature in Australian defense and national security planning and maritime strategy," and that the ADF needs to be prepared to play a part in securing these sea lanes.[16] This is consistent with Australia's Indian Ocean littoral status and the increased prominence of the Indian Ocean in developing U.S. strategic policy. While the emphasis given to the Indo-Pacific region is in some respects a continuation of previous defense thinking, the increased focus is significant nonetheless.

The final noteworthy thematic departure of the 2013 white paper is the emphasis on fiscal uncertainty. The chapter on finances commits the government to a defense budget that delivers the capabilities to meet preparedness requirements and to protect Australia's national security interests. But it flags that the Australian fiscal environment "remains challenging" and stipulates that this commitment is subordinate to the priority the government places on improving the overall budget situation.[17] Many Australian defense commentators expressed skepticism about the likelihood of the force structure outlined in the white paper being adequately funded and about the likelihood that defense spending would return to the aspirational target of 2 percent of GDP.[18]

AUSTRALIA'S DEFENSE BUDGET: NOT KEEPING UP WITH THE JONESES

After the 10 percent 2012–13 defense budget cut, the 2013–14 budget represented something of a return to normalcy. The government allocated AUD25.3 billion for defense, an increase of AUD1.2 billion (2.25 percent) over the previous year. This modest increase will nudge defense spending from 1.56 percent of GDP to 1.6 percent (see Figure 3-1). The planned allocation will grow to AUD30.7 billion in 2016–17, with AUD 8.3 billion budgeted for new projects across the next 4 years — representing real growth of 10 percent annually in capital investment.[19]

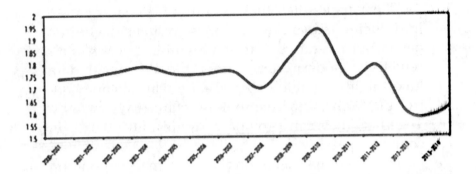

Source: "The Cost of Defense: ASPI Defense Budget Brief, 2013-14," Barton, Australia: Australian Strategic Policy Institute, May 30, 2013, available from *www.aspi.org.au/publication/publications_all.aspx*.

Figure 3-1. Defense Expenditure as a Percentage of GDP, 2000-14.

Overall, however, the budget does not redress the cuts of the previous 2 years.[20] The budget documents reaffirm the government's intention to attain the 2 percent of GDP target, but this will not happen soon: "This is a long-term objective that will be implemented in an economically responsible manner as and when fiscal circumstances allow."[21] According to the government's own estimates, defense spending will be capped as a share of GDP at 1.66 percent through at least 2017–18.[22]

This leisurely return to a credible level of defense spending is difficult to reconcile with a regional security environment that, if anything, has deteriorated since the 2009 white paper was published. As the 2013 paper makes clear: "We are witnessing the evolution of a more complex and competitive order" and "Australia's relative strategic weight will be challenged as the major Asian states continue to grow their economies and modernize their military forces."[23]

The paper notes that neighboring countries are introducing advanced weapons systems including beyond-visual-range air-to-air missiles, air-to-air refueling, modern surveillance radars, digital data links, highly capable airborne early warning and control platforms, and electronic warfare (EW) systems. Together, they can provide a significant increase in combat capability.[24]

For Australia—a country with a small population that occupies a vast island continent with an extensive coastline and massive territorial waters—maintaining sophisticated forces with a technological edge over neighboring countries has long been a keystone of its defense policy. Recent regional defense acquisition trends are reducing the strategic depth that has long benefited Australian security and are making it more expensive to maintain that capability edge.

As noted previously, the 2013–14 budget does commit additional funds for procuring equipment, including AUD2.94 billion to acquire 12 EA-18G Growler aircraft, which will complement the existing 24 F/A-18E/F Super Hornets purchased in 2006, as a hedge against late delivery of the F-35 Lightning II aircraft.[25] Australia has allocated up to AUD16 billion for the F-35 program, with plans to buy up to 72 F-35s initially and potentially another 28 later on. The budget also allows for:

- Fast-tracking replacement vessels for the existing fleet of Armidale-class patrol boats;
- Replacing two fleet replenishment ships;
- Installing Australian-designed phased-array radar on the navy's future frigates;
- Establishing a joint U.S.–Australia-operated C-band radar space surveillance installation in Western Australia; and,

- Acquiring P-8 Poseidon maritime surveillance aircraft.[26]

Funding for operations is down, ostensibly reflecting the drawdown of ADF operations in Australia's immediate region and the departure of 1,000 of Australia's 1,650 troops in Afghanistan by the end of 2013. Spending on operations will drop from AUD1.5 billion in 2012–13 to less than AUD1.0 billion in 2013–14.

Australia's two largest political parties—Labor and Liberal—agree that current levels of defense funding are inadequate. As noted earlier, the current Labor government has set 2 percent of GDP as a long-term goal for defense spending, conditioned on the fiscal situation. The Liberal Party, which leads the Coalition of opposition parties, has said it will "restore sensible defense spending to 3 percent real growth per year subject to improvements in the Budget."[27] In both cases, the devil will be in the details. The outlook is far from promising, owing to rapid expansion in government spending by successive Labor governments, ballooning health care costs, and deteriorating national revenues tied to a variety of factors, including a slowdown in commodity exports to China. As a result, economic forecasters are warning that Australia could face annual budget deficits for the next decade.[28] Add in the costs of new entitlement programs, and it is difficult to be optimistic about Australia returning to a credible level of defense spending anytime soon.

This funding crunch comes at a time when the United States faces its own fiscal pressures and deep defense budget cuts. Consequently, Washington's expectations for its allies are rising. The Obama administration is demanding U.S. partners make credible contributions to defense and security by maintaining modernized and ready forces and by taking the lead in regional security and stabilization operations. Senator John McCain's pointed criticism of the Gillard government's defense cuts likewise suggests that a future Republican administration is unlikely to have lower expectations.[29] Australia has made a clear commitment to support the U.S. rebalance to the Asia-Pacific, yet it remains to be seen whether the Australian defense budget will be able to meet its side of the capability, readiness, and operational bargain.

PROCUREMENT PROGRAMS: SEEKING AN EDGE

Australia has introduced a number of significant new military capabilities in the past decade, many of them as a result of decisions made by the Howard government. These include E-7A Wedgetail airborne early warning aircraft, C-17 Globemaster III transport aircraft, KC-30 air-to-air refuelers, M1 Abrams tanks, F/A-18E/F Super Hornet combat aircraft, and Tiger ARH attack helicopters. The regular army and special forces have expanded since 2000. Nevertheless, Australia faces a number of major capability challenges over the next decade and beyond.

Foremost among these is replacing Australia's increasingly unreliable fleet of six conventional Collins-class submarines. These boats have been plagued with problems since they were delivered between 1996 and 2003, including propulsion system issues, poor availability, a shortage in skilled operators, and significant limitations in combat capability.[30] Efforts to address some of these problems with a new combat system and the acquisition of new heavyweight torpedoes began in 2002. But based on current plans, it will not be until 2016 that all of the submarines will have the new combat system installed.[31] Moreover, retirement of the Collins-class fleet is expected between 2022 and 2031, resulting in a potential submarine capability gap in the late-2020s.[32]

The 2009 white paper committed the government to acquiring 12 larger, more capable conventional submarines to replace the Collins class—all of which were to be built in South Australia. Despite many commentators' views that this commitment was financially unsustainable and technologically beyond Australia's reach, the 2013 white paper reaffirmed both the need for 12 conventionally powered submarines and the plan to have them built in Australia.[33]

Australia's long coastline and distances from key operating areas necessitate a submarine force with extensive range and endurance, capabilities that go well beyond those provided by most conventional designs. These characteristics can only be incorporated in a very large hull. Indeed, a number of Australian and U.S. analysts have argued that Australia's needs could be best met by acquiring nuclear-powered submarines, most likely from the United States.[34]

Despite arguments in favor of this option, it remains politically controversial and the Labor Party has expressly ruled it out. Officials have, however, confirmed that the submarines will be equipped with U.S. heavyweight torpedoes and a U.S.-supplied combat system.[35] Defense technological cooperation with Japan on this front is also a possibility.[36] But in the meantime, the looming submarine capability gap is becoming a matter of increasing urgency for Australia's defense planners.

The other major potential gap is in air-combat capability. Australian governments are typically sensitive to any suggestion of a gap in air capabilities. Faced with an aging F/A-18A/B fleet, the earlier-than-anticipated withdrawal from service of its fleet of F-111 Aardvark strike aircraft, and delays in the development of the fifth-generation F-35, the Howard government decided to purchase 24 Super Hornets as a hedge against this eventuality.

The 2013 white paper continues this prudential approach. It takes note of the emerging advanced air-combat and air-defense systems in the region, the proliferation of modern EW systems, and the growing risk EW systems pose to Australia's ability to control the air, conduct strikes, and support land and naval forces.[37] Against this challenging backdrop, the white paper is unequivocal that "The Government will not allow a gap in our air-combat capability to occur."[38] It reaffirms Australia's commitment to the F-35 program, with an expectation that three operational squadrons of up to 72 aircraft will enter service around 2020. In response to the proliferation of sophisticated EW systems, it also commits to acquiring 12 new Growler electronic attack aircraft, which will make Australia the only country outside of the United States with this capability.

Together with the six E-7 Wedgetail early warning and control aircraft, these new systems will provide Australia with significantly enhanced networking capability among its forces, interoperability with U.S. forces, and the ability to operate in a more "contested" regional environment. After delays in development, the Wedgetail is now meeting or exceeding performance parameters and will have the capability to detect and identify potential enemy electronic emissions at great ranges.[39] The Royal Australian Air Force's future dependence on the F-35, however, means that

Canberra will remain acutely sensitive to any further delays and capability issues affecting the program and to future reductions in the overall size of the program that would drive up the F-35's unit cost.

The middle of this decade will also see the transformation of Australia's amphibious capabilities with the introduction into service of two Spanish-designed Landing Helicopter Docks (LHDs) which, at 27,000 tons, will be the largest-ever ships to serve with the Royal Australian Navy. They will improve interoperability with the United States and regional partners and increase Australia's ability to respond to a range of contingencies. The rotation of U.S. Marines in Northern Australia will provide extensive training opportunities to build on Australia's increased amphibious capabilities.

Under Plan Beersheba, the Australian Army is being restructured into three multirole combat brigades, including a battalion designated as the core of a future amphibious force.[40] It remains unclear, however, how much ground combat power Australia will be able to deploy and sustain. A combination of capability and political considerations has constrained the situations in which the Australian government has been prepared to use land forces.

In 2006, for example, Canberra deployed an amphibious task force to waters off of Fiji in response to an anticipated military coup. However, a major factor in the government's ultimate decision to not intervene was the concern that the ADF lacked the firepower to overcome the well-trained Fijian military at an acceptable cost to Australian forces. In Iraq and Afghanistan, the government preferred to commit special forces to initial combat operations rather than commit larger ground forces, again in part because of perceived capability limitations in firepower, force protection, and combat enablers and the resulting political risks such a deployment would entail.

In both cases, Australia did subsequently take on larger stabilization responsibilities—in Iraq's Al Muthanna Province and in Afghanistan's Uruzgan Province. It is unclear, however, whether the ADF could have on its own held down a "hotter" province in either country for a prolonged period if the Australian government had made a decision to do so—as the Australian Army did in Vietnam, for example.

As Australian force planners examine the lessons learned from recent operations, they should advise Australia's current and future political leaders on whether the ADF has the capabilities they assume it has and, if not, whether the ADF should develop them. Coalition planners in the Pentagon likewise need to know what the ADF's actual capabilities are.

The 2013 white paper left two other key capability decisions unresolved. The first of these is whether to equip Australia's three new air-warfare destroyers with Standard Missile 3s so Australia can be involved in missile defense operations. The ships will be actively outfitted with the U.S. Aegis Combat System, capable of detecting and tracking a variety of missiles including ballistic missiles, and will operate with American, Japanese, and South Korean naval forces. While the white paper recognizes the increasing threat posed by ballistic missiles, it rather vaguely commits the government to "continue to examine potential Australian capability responses."[41]

The second unresolved matter is cruise missiles. Currently, Australia's main weapon for strike missions is the Joint Air-to-Surface Standoff Missile, launched from the air force's F/A-18 aircraft and with a range of over 200 nautical miles. This capability will be augmented when the F-35 is introduced, with its stealth characteristics and suite of precision weapons and ISR systems. The 2009 white paper, however, went further, committing the government — in a major departure for Australia and for the Southeast Asian region — to acquiring maritime-based land-attack cruise missiles to be fitted to the new air-warfare destroyers, future frigates, and to the successors to the Collins-class submarine fleet.[42]

With no explanation, however, the 2013 white paper seems to have stepped back from this commitment, stating only that it would look into "options for the Government to expand strategic strike capabilities if required."[43] This development presumably owes as much to the government's current fiscal problems as it does to any alteration in the regional security environment outlined in the 2009 white paper.

SUSTAINABILITY, READINESS, AND POSTURE IN NORTHERN AUSTRALIA

Sustainment has been a major challenge for the ADF since the late-1990s. Multiple operations abroad have placed significant strain on personnel, equipment, and support systems. ADF recruitment and retention have generally held up well, with Australia's military forces enjoying public support and with enhanced pay and housing conditions boosting the attractiveness of military service. With the acquisition of C-17 and C-130J Super Hercules transport aircraft and the LHDs, the ADF will enjoy enhanced strategic lift capabilities and an increased capacity to support deployed forces.

Although the defense budget has come under pressure in the last few years, the 2013 white paper avoided declaring a post-drawdown "peace dividend," stating that despite the more fiscally constrained environment, there would be no reduction in overall ADF personnel numbers.[44] Significant pressures and deficiencies remain, however. Shortages of specialist skills in some areas have been exacerbated by the Australian minerals boom, with the demand for engineering and related trades in particular draining individuals from the military.

This has resulted in reduced operational availability in some arms of the ADF such as the submarine fleet. The navy's amphibious fleet has suffered a series of major mechanical failures owing to systemic sustainment and maintenance failures, forcing the government to make the rushed purchase of a former British vessel to make up the shortfall.[45] As noted earlier, the army's fleet of light armored vehicles has experienced unanticipated wear and tear as a result of sustained deployments and will require significant rehabilitation as troops deploy back to home bases.

Maintaining readiness during a period of reduced operational tempo will be another major challenge for the ADF. One possible consequence of the reduction in the operational budget noted previously will be fewer funds for training.[46] This is likely to affect the active duty army in particular, but will also impact the training for reserve forces.

A more uncertain regional security environment, the growing strategic importance of the Indian Ocean, and community concerns about the potential vulnerability of Australia's vital natural resources led the government to commission a review of the ADF's force posture in 2011.[47] The review found that the

ADF needs to be postured to support high-tempo military operations in Australia's northern and western approaches and recommended a number of steps to strengthen the ADF's presence and ability to sustain such operations, including:

- Upgrading airbases in Northern and North Western Australia to handle larger aircraft types (necessary to implement the agreement reached during President Obama's November 2011 visit to Australia for increased rotations of U.S. aircraft);
- Increasing ADF aircraft and ship deployments to the area;
- Upgrading airfield facilities at Cocos Island (an offshore Australian territory proximate to the Bay of Bengal and the western approaches to the strategically vital Strait of Malacca) to support future operations by P-8A maritime surveillance aircraft and unmanned aerial vehicles (UAVs);
- Expanding facilities at HMAS Stirling, the Royal Australian Navy's major west-coast base near Perth, to support deployments by major surface combatants of the U.S. Navy;
- Giving consideration to hardening forward-operating bases; and,
- Enhancing facilities and opportunities for training with U.S. and other partner militaries.

The government has accepted the thrust of the force posture review and is already implementing some of its more straightforward recommendations.[48] The government also announced that it would seek opportunities with the United States to fund jointly improvements to bases, facilities, and training infrastructure as part of the enhanced practical defense co-operation measures announced in 2011.

THE U.S.-AUSTRALIA ALLIANCE: SOUTHERN HINGE OF THE U.S. PIVOT?

U.S.-Australia security cooperation deepened and broadened significantly during the post-9/11 decade. This included closer operational, intelligence, and counterterrorism collaboration; greater Australian access to U.S. defense information and systems; and the 2007 signing of the Australia–U.S. Defence Trade Cooperation Treaty to streamline defense industrial cooperation. The treaty, which came into force in May 2013, is intended to facilitate exports of defense goods, services, and technology and to improve delivery times and sustainment. It complements the ADF's acquisition of a range of weapons systems that are able to operate seamlessly with U.S. forces.

Initial talks on enhanced U.S. military access to Australia preceded the Obama administration's pivot, or rebalance, to Asia, and should be seen in the context of intensifying strategic links. The talks were quietly initiated by the Howard government in 2007 with the George W. Bush administration, building not only on the post-9/11 alliance relationship but also with an eye toward shifting power dynamics in the Asia-Pacific region.[49]

By 2011, governments around the region were becoming concerned with China's increasingly assertive behavior and looking for reassurance about America's staying power in the Western Pacific following the 2008 global financial crisis. Washington, for its part, was seeking options to facilitate a more distributed military footprint in Asia, closer engagement with

Southeast Asia, and enhanced access to vital Indian Ocean sea lanes of communication.

The result was President Obama's speech to the Australian parliament in November 2011 in which he laid out Washington's rebalancing strategy. During his visit, the two governments announced that Australia was the first country in the region to agree to an enhanced U.S. military presence. This would include both a rotational Marine Corps presence in Darwin — which, by 2016–17, would build to a 2,500-strong Marine Air-Ground Task Force — and increased use by U.S. Air Force aircraft of airbases in Northern Australia.

The current marine rotation numbers around 200, and an assessment has just been released to prepare for rotations of up to 1,100 personnel.[50] While there has been some concern that the Labor Party's support for these initiatives may be ebbing, Defense Minister Stephen Smith is on record as stating that the government's current fiscal difficulties will not have an adverse impact on enhanced cooperation with the United States.[51]

What could have an impact over time, however, is an increasingly vocal strand of elite opinion in Australia that sees the country's growing economic interdependence with China as incompatible with its security ties to the United States.[52] Beijing exploits this anxiety in an increasingly sophisticated public diplomacy effort in Australia.[53] U.S. officials, however, have grounds for cautious optimism on this score. First, neither major political party shows any sign of a weakened commitment to the U.S. alliance. The Labor Party went out of its way to state in the 2013 white paper that "The Government does not believe that Australia must choose between its long-standing Alliance with the United States and its expanding relationship with China."[54] The Liberal-led coalition took this view under Howard and maintains that position.

Second, public opinion is unequivocal. Support for the U.S. alliance is strong, with more than 80 percent of Australians regarding the alliance as either "very important" or "fairly important" for Australia's security.[55] Nearly three-quarters believe the United States will be Australia's most important security partner over the next decade and a similar proportion are in favor of "up to 2,500 U.S. soldiers being based in Darwin."[56] Third, China's behavior in disputed waters in the Western Pacific shows few signs of moderating and is likely to sustain support for the alliance as a hedge against future uncertainty and as a counterweight to China's economic influence in Australia. (See Figure 3-2.)

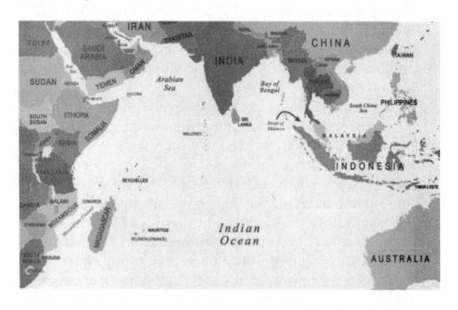

Figure 3-2. Australia's Strategic Neighborhood.

Assuming that the base of support for the alliance in Australia remains strong, enhanced defense cooperation is likely to deepen and to continue extending into newer areas. During World War II, Australia was vital to the U.S. Pacific War effort because Australia offered strategic depth and access to crucial sea lanes of communication. The joint Australia–U.S. intelligence facilities have made a vital contribution to the security of both nations for the past several decades. While circumstances are obviously different today, those strategic considerations remain important.

In addition to the aforementioned increased deployments of American sea and air assets to Australia, the ramped-up program of amphibious training exercises with the ADF, and the potential creation of a genuinely joint expeditionary capability, discussions have started about the use of the enhanced airfield on Cocos Island to support operations by maritime surveillance aircraft and UAVs. Australia and the United States have also stepped up defense cooperation in the realms of cyberspace and space, including the establishment of the new joint space surveillance installation in Western Australia.

Close defense industrial, intelligence, and operational cooperation will also remain vitally important, particularly for Australia. U.S. technical support was essential to rehabilitating the Collins-class submarines, and the 2013 white paper makes explicit that Canberra will look to Washington for assistance to deliver its ambitious submarine replacement program.[57] Integrating the sophisticated F-35 into the ADF and networking it with a suite of other interoperable capabilities will likewise require unprecedented levels of Australia–U.S. collaboration.

The ADF can make an important operational contribution to the evolving Air-Sea Battle (ASB) concept, not only by facilitating a more distributed American force posture in the Asia-Pacific region, but also by participating in "distant blockade" operations around Southeast Asian maritime chokepoints and by augmenting U.S. enabling capabilities such as tanker aircraft, EW assets, and strategic lift in contingencies.[58]

But deepening cooperation will require greater effort on Washington's part to articulate more fully its vision for ASB and the roles allies are expected to play. At the same time, deepened defense and strategic ties will require more maturity in Australia about the need for closer involvement in a range of detailed U.S. military contingency planning and the attendant diplomatic challenges that will inevitably arise as a result.

CONCLUSION

Australia, the United States, and their alliance face major strategic challenges as global power shifts increasingly toward Asia. China's rise and its rapid military modernization are transforming the regional security environment. The People's Liberation Army's development of anti-access and area denial capabilities is challenging the U.S. military's ability to operate in the Western Pacific and is reshaping the regional military balance. Moves by other Asian powers to acquire sophisticated weapons are contributing to a more complex and contested region and eroding Australia's long-standing military capability edge.

Against this backdrop, it seems imperative that Australia's own military modernization agenda proceed apace. As a result of the Gillard government's defense cuts and predicted revenue pressures for the next decade, however, Australia's modernization plans are now at risk. The funding shortfalls outlined in the 2009 and 2013 white papers may be as much as AUD33 billion for the period 2009–22.[59] The consequence, according to a leading expert on the Australian defense budget, is an inevitably slow modernization of the defense force.[60]

A related alliance-based challenge will be managing U.S. expectations. Reasonably enough, the United States is looking to its Asia-Pacific allies, including Australia, to shoulder a greater share of the burden of maintaining a favorable balance of power in the region. Continuing support for enhanced defense cooperation in Australia is part of this expectation, as is an increased Australian contribution to maintaining deterrence through stepped-up operational cooperation. Australia is doing this unobtrusively in the realms of space and cyber warfare, intelligence collection, and ballistic missile early warning, but it must accept a more public and upfront role in other areas such as missile defense and participation in ASB.

Australia will also need to continue efforts to step up its own defense engagement with other U.S. regional partners such as Japan, Indonesia, India, and South Korea, utilizing mechanisms such as the Australia–Japan–U.S. Trilateral Strategic Dialogue and establishing new, informal "minilateral" security groupings that incorporate India and Indonesia. It is unclear how the forthcoming Australian election will affect the nation's defense policy. The center-right coalition traditionally places importance on defense, and the opposition has committed itself to producing a new, properly priced defense white paper within 18 months of taking office and to making the necessary decisions within that time frame to avoid any submarine capability gap.[61]

The opposition has also signaled a less constrained approach to supporting U.S. military forces in Australia should it win office. Ultimately, however, the Liberal Party-led coalition's ability to deliver on defense would depend on its success in restoring the budget to a sustainable trajectory and the priority it places on defense and maintaining a strong U.S. alliance.

The defense implications of a Rudd election victory are even less clear. Judging by the 2009 white paper, Rudd's instincts on defense are hawkish, and he may seek to restore its ambitious force structure goals. His ability to deliver on them, however, would be significantly constrained by Australia's difficult fiscal outlook and his own party's appetite for increased domestic spending.

The jury will remain out until Australia's new government confronts its own inevitable first national security test and delivers its first defense budget.

ENDNOTES - CHAPTER 3

1. This chapter was originally published as an essay on August 22, 2013.

2. Department of Defence, White Paper: Defence 2000, Canberra, Australia: Australian Government, 2000, p. xvii.

3. See, for example, Greg Sheridan, *The Partnership: The Inside Story of the US-Australian Alliance under Bush and Howard,* Sydney, Australia: University of New South Wales, 2006.

4. Department of Defence, Defence White Paper 2009, *Defending Australia in the Asia Pacific Century: Force 2030,* Canberra, Australia: Australian Government, 2000.

5. WikiLeaks cables subsequently revealed that earlier in 2009, Rudd told incoming U.S. Secretary of State Hillary Clinton that Australia's intelligence agencies were closely monitoring China's military expansion and that Australia's planned naval build-up was in direct response to China's growing power-projection capabilities. See Dan Flitton, "Rudd the Butt of WikiLeaks Expose," *The Age*, December 6, 2010, available from *www.theage.com.au/technology/security/rudd-the-butt-of-wikileaks-expos-20101205-18lf2.html*.

6. Andrew Shearer, "Australia Bulks Up," *The Wall Street Journal Asia*, May 6, 2009, p. 15.

7. *The Cost of Defence: ASPI Defence Budget Brief 2012–13*, Canberra, Australia: Australian Strategic Policy Institute, 2012, p. vi.

8. Department of Defence, Defence White Paper 2013, Canberra, Australia: Australian Government, 2013, p. 11.

9. *Ibid.*, pp. 9, 11.

10. *Ibid.*, p. 9.

11. *Ibid.*, pp. 61–62.

12. John Garnaut, "Why the World is Reading Gillard's Defence Paper," *The Age*, May 10, 2013, available from *www.theage.com.au/federal-politics/political-opinion/why-the-world-is-reading-gillards-defence-paper-20130509-2jak2.html*.

13. On the latter point, see Andrew Shearer, *Sweet and Sour: Australian Public Attitudes toward China*, Sydney, Australia: Lowy Institute for International Policy, August 1, 2010, available from *www.lowyinstitute.org/publications/sweet-and-sour-australian-public-attitudes-towards-china*.

14. Julia Gillard, "Australia's National Security Beyond the 9/11 Decade," speech, Canberra, Australia: Crawford School of Public Policy, Australian National University College of Asia & the Pacific, January 23, 2013, available from *https://acbee.crawford.anu.edu.au/news/132/prime-minister-julia-gillards-speech-national-security-college*. The 2013 white paper acknowledged that, in the

period following the East Timor intervention and 9/11, the Australian military has experienced a high operating tempo, with the ADF having undertaken some 100 operations since 1999. It also correctly highlights the importance of learning the right lessons from the current period and avoiding the pitfalls that followed the ADF's withdrawal from Vietnam in the 1970s, such as the loss of hard-won counterinsurgency capabilities.

15. Defence White Paper 2013.

16. *Ibid.*, pp. 25–26.

17. *Ibid.*, p. 71.

18. See, for example, Paul Dibb, "Show Us the Money for Defence Spending," *The Australian*, May 6, 2013, available from *www. theaustralian.com.au/national-affairs/opinion/show-us-the-money-for-defence-spending/story-e6frgd0x-1226635611440*.

19. Mark Thomson, "Second Chance: Will They Deliver?" Defence Special Report, May 25–26, 2013, p. 1, available from *www.aspi.org.au/publications/opinion-pieces/second-chance-will-they-deliver/Second_chance-will_they_deliver.pdf*.

20. *Ibid.*

21. See John Kerin, "Cuts Halted Despite Troop Drawdown," *Australian Financial Review*, May 15, 2013, available from *www.afr. com/p/national/budget/cuts_halted_despite_troop_drawdown_8bYul1v RFZka5t7GJYN6jI*.

22. See Christopher Joye, "The Lowest Military Spending Since 1938," *Australian Financial Review*, May 15, 2013, available from *www.afr.com/p/national/budget/the_lowest_defence_spending_ since_aaGRxTIKT1YZFOZzYfneBK*.

23. Defence White Paper 2013, pp. 8, 15.

24. *Ibid.*, p. 14.

25. Australian Government, Department of Defence, Budget: Portfolio Budget Statements 2013–14: Budget Related Paper No. 1.4A: Defence Portfolio: Budget Initiatives and Explanations of

Appropriations Specified by Outcomes and Programs by Agency, Canberra, Australia, 2013.

26. Stephen Smith, "Minister for Defence—Budget 2013–14: Defence Budget Overview," May 14, 2013, available from *www. minister.defence.gov.au/2013/05/14/minister-for-defence-budget-2013-14-defence-budget-overview/*.

27. *Our Plan—Real Solutions for All Australians: The Direction, Values and Policy Priorities of the Next Coalition Government*, Canberra, Australia: Liberal Party of Australia, 2013, p. 48.

28. See David Crowe and David Uren, "Treasury Rings Alarm on Surplus as Budget Hole Means Decade of Deficits," *The Australian*, May 23, 2013, available from *www.theaustralian.com. au/national-affairs/treasury-rings-alarm-on-surplus-as-budget-hole-means-decade-of-deficits/story-fnhi8df6-1226648770308*.

29. See Greg Sheridan, "McCain Slams Labor for 'Imprudent' Cuts," *The Australian*, April 27, 2013, available from *www. theaustralian.com.au/national-affairs/defence/mccain-slams-labor-for-imprudent-cuts/story-e6frg8yo-1226630344660*.

30. See Andrew Davies and Mark Thomson, "Mind the Gap: Getting Serious about Submarines," *Strategic Insight,* No. 57, Canberra, Australia: Australian Strategic Policy Institute, April 2012, pp. 2–6.

31. Ibid., p. 6.

32. The white paper suggests, perhaps optimistically, that it might be possible to close this gap by extending the retirement window by 7 years. See Defence White Paper 2013, p. 83.

33. Smith, "Minister for Defence—Budget 2013–14."

34. See, for example, Ross Babbage, "Why Australia Needs Nuclear Subs," *The Diplomat*, November 8, 2011, available from *thediplomat.com/2011/11/08/why-australia-needs-nuclear-subs/*; and John Kerin and Christopher Joye, "Labor Split on Nuclear Submarines," *Australian Financial Review*, November 12, 2012, available from *www.afr.com/p/national/labor_split_on_nuclear_submarines_Q5hmLHBo7pneF4K8LpZ4wN*.

35. Brendan Nicholson, "Subs Need To Be Out There Doing the Damage," *The Weekend Australian*, May 25–26, 2013.

36. Japan is building its own advanced 4,000-4,200 ton submarine. See, for example, Rex Patrick, "Japanese Flavoured Submarines for Sea 1000," *Asia-Pacific Defence Reporter*, October 31, 2012, available from *www.asiapacificdefencereporter.com/articles/270/JAP-ANESE-FLAVOURED-SUBMARINES-FOR-SEA-1000*.

37. Defence White Paper 2013, p. 88.

38. *Ibid.*

39. Kym Bergmann, "Better Late Than Never: Wedgetails Overcome Obstacles to Become Envy of the World," *Defence Special Report*, May 25–26, 2013.

40. Defence White Paper 2013, p. 85.

41. *Ibid.*, pp. 81–82.

42. Defence White Paper 2009, p. 81.

43. Defence White Paper 2013, p. 77.

44. *Ibid.*, p. 71.

45. The vessel is the HMAS Choules, formerly Royal Fleet Auxiliary Largs Bay. For details on sustainment and maintenance issues affecting the Royal Australian Navy's amphibious fleet, see Paul J. Rizzo, *Plan to Reform Support Ship Repair and Management Practices*, Canberra, Australia: Australian Government, Department of Defence, 2011.

46. See, for example, John Kerin, "Military training to get chop in budget," *Australian Financial Review*, April 2, 2013, available from *www.afr.com/news/policy/defence/military-training-to-get-chop-in-budget-20130401-j0yo5*.

47. Allan Hawke and Ric Smith, *Australian Defence Force Posture Review*, Canberra, Australia: Australian Government, Department of Defence, March 30, 2012.

48. Defence White Paper 2013, pp. 47–51. The paper announced, for example, that plans will continue to enhance HMAS Stirling to support major surface combatant and submarines operations, and it also confirmed the upgrading of the Cocos Island airfield.

49. I was former Prime Minister Howard's international policy adviser at the time.

50. Brendan Nicholson, "US Alarm At 'Cooling' On Marines," *The Australian*, April 2, 2013, available from *www.theaustralian. com.au/national-affairs/defence/us-alarm-at-cooling-on-marines/story-e6frg8yo-1226610490588*; and Australian Associated Press, "200 More US Marines To Arrive in Darwin," April 20, 2013, available from *www.news.com.au/breaking-news/national/more-us-marines-to-arrive-in-darwin/story-e6frfku9-1226624990044*.

51. Smith, "Minister for Defence—Budget 2013–14."

52. Hugh White has been the leading academic advocate of this argument. See Hugh White, *Power Shift: Australia's Future between Washington and Beijing*, Melbourne, Australia: Black Inc., September 2010. Others expressing misgivings include former Labor Prime Minister Paul Keating, former Coalition Prime Minister Malcolm Fraser, former Coalition leader and current opposition front-bencher Malcolm Turnbull, and leading business figure Kerry Stokes, who said he was "physically repulsed" by the thought of U.S. forces on Australian soil. See Phillip Hudson, "Stop Lecturing China, Says Stokes," *Herald Sun*, September 14, 2012, available from *www.heraldsun.com.au/news/national/stop-lecturing-china-says-stokes/story-fndo48ca-1226474468140*.

53. This reportedly includes efforts by Chinese intelligence to influence elite opinion. See, for example, John Garnaut, "Chinese Spies Woo Business Leaders," *The Age*, May 25, 2013, available from *www.theage.com.au/business/world-business/chinese-spies-woo-business-leaders-20130524-2k717.html*.

54. Defence White Paper 2013, p. 11.

55. Fifty-four percent say "very important," 28 percent say "fairly important." See Alex Oliver, *The Lowy Institute Poll 2013:*

Australia in the World: Public Opinion and Foreign Policy, Sydney, Australia: Lowy Institute for International Policy, June 24, 2013, p. 7. See also Fergus Hanson, *The Lowy Institute Poll 2012: Australia and New Zealand in the World: Public Opinion and Foreign Policy*, Sydney, Australia: Lowy Institute for International Policy, June 5, 2012, pp. 9–10.

56. Support for basing U.S. military forces in Australia has grown from 55 percent in 2011 to 61 percent in 2013, perhaps in response to increasing reports of China's military assertiveness during 2012. See Olivier, The Lowy Institute Poll 2013, p. 8.

57. Defence White Paper 2013, p. 82.

58. See Ben Schreer, "Australia and Air-Sea Battle," *PacNet Newsletter*, No. 30, May 1, 2013.

59. John Kerin, "Arms Wish-List $33bn Short," *Australian Financial Review*, May 30, 2013. At least one senior Obama administration official has voiced clear criticism of the Gillard government's defense cuts. See Peter Hartcher, "US To Take Up Defence 'Freeloading' with Cabinet," *Sydney Morning Herald*, November 10, 2012, available from *www.smh.com.au/federal-politics/political-news/us-to-take-up-defence-freeloading-with-cabinet-20121109-293dr.html*.

60. Thomson, "Second Chance: Will They Deliver?"

61. *Our Plan — Real Solutions for All Australians*, p. 48.

CHAPTER 4

THE NORTH ATLANTIC TREATY ORGANIZATION AT SEA: TRENDS IN ALLIED NAVAL POWER[1]

Bryan McGrath

KEY POINTS

- The North Atlantic Treaty Organization's (NATO) intervention in Libya during the spring and summer of 2011 raised serious questions about the naval capabilities of America's European allies.
- Despite declining defense budgets, the major European naval powers have sought to retain a broad array of naval capabilities, resulting in modern but substantially smaller fleets.
- With U.S. armed forces increasingly focused on the Asia-Pacific region, there are growing concerns as to whether the navies of America's continental allies are up to meeting the challenges arising from the general unrest on Europe's eastern and southern maritime flanks.

Taking its name from one of the world's great oceans, NATO has throughout its history been a military alliance focused primarily on land. Although several of its members have built and maintained first-rate navies, sea power served largely as a flanking force for what was envisioned as the main Cold War battle on the central front. After the fall of the Soviet Union, land conflict continued to be a primary emphasis of the alliance, first in dealing with the disintegration of

Yugoslavia, and then as NATO assumed a central role in the Afghan conflict.

That said, naval power has historically been a defining feature of the alliance. While the United States provided a preponderance of alliance naval power, several allies—including the United Kingdom (UK), France, Germany, Spain, and Italy—created fleets capable of global power projection, and others chose to pursue niche capabilities to supplement the striking power of the larger fleets. This chapter assesses the state of the former group.

It is a propitious time to review NATO's naval capabilities. Continental Europe is at peace. The only trouble has been on NATO's eastern and southern maritime flanks. Unrest throughout North Africa and the Levant raises the very real possibility that NATO's European nations will have to shoulder a larger share of a growing maritime security burden than they have been accustomed to or have been preparing for. The largest naval force contributor to the alliance—the United States—is increasingly focusing its attention on the Pacific, and it has not routinely operated large naval task forces in the Mediterranean Sea for decades. The 2011 military intervention in Libya and recent discussions about possible intervention in the Syrian civil war raise questions about NATO's ability to project naval power effectively, especially without the full participation of the U.S. Navy.

Several trends are evident among the major NATO navies. First, they are getting smaller. All of the navies analyzed here have fewer ships today than in the year 2000—in some cases, significantly fewer. While ship counts do not tell the entire story of a nation's naval might (especially in the age of networked operations), they remain a useful proxy for naval capability, especially with respect to blue-water operations far from home waters. The primary reason these navies are getting smaller is a decline in general defense spending, including shipbuilding.

Second, the ships that are being built are increasingly capable and sophisticated—and therefore expensive—which serves only to drive down fleet size in an era of fiscal restraint.

Third, historically maritime nations seem to desire to retain broad, general purpose fleets even if it means smaller fleets overall. For example, the once-mighty UK Royal Navy is planning for a surface fleet of only 19 major surface combatants, while moving forward on construction of two aircraft carriers and a replacement submarine class for its aging strategic deterrent, both of which consume considerable shipbuilding resources.[2]

OPERATION UNIFIED PROTECTOR

The controversy over the participation of major NATO partners in the Libyan intervention has encompassed operational effectiveness as well as political will. The contributions of the five major allies surveyed in this chapter vary widely. Britain and France proved both highly capable and highly committed, while Italy, Spain, and Germany provided, respectively, partial, minimal, and nonoperational support.

NATO's reliance on the United States from March to October 2011 to carry out the allied mission—despite President Barack Obama's admonition that the United States would not take the lead in the military operation—is the result of two distinct causes: NATO-wide underinvestment in military capability and a lack of political will on the part of uniquely capable countries. Capability is absent in some areas; in others, it is unevenly distributed. When key platforms were present and fielded, they were often numerically too few.

The case of the *Charles de Gaulle* demonstrates that numbers matter. France's aircraft carrier, the only non–U.S. catapult assisted take-off but arrested recovery carrier in Europe, accounted for 33 percent of allied-strike sorties before its withdrawal in August 2011.[3] The endurance of even the largest ships is limited, however, by crew fatigue and maintenance requirements.[4] When Italy, citing austerity measures, withdrew its carrier, the *Giuseppe Garibaldi*, from the Libyan operation, only amphibious ships and short take-off and vertical landing (STOVL) carriers remained to replace the *de Gaulle*.[5] Indeed, extended global deployments preceding those to Libya taxed even the endurance of U.S. amphibious warships, which departed before *de Gaulle* and *Garibaldi*.[6] The remaining large-deck ships—the French landing platform dock (LPD) *Tonnerre* and the British vessels *Albion* and *Ocean*—supported only attack helicopters.[7] Lack of available land-based aircraft would have resulted in significantly slower operations.

Operation UNIFIED PROTECTOR, the NATO name given to the Libyan campaign, cannot be considered a stressing scenario for NATO's naval and air forces. Targets were located primarily along Libya's coast, well within the range of land-based aircraft. The enemy was entirely unprepared for NATO intervention.[8] The strategic geography of the Libyan civil war greatly facilitated intervention. To attack rebel-held areas, Muammar Gaddafi's forces often had to move across long stretches of flat, exposed, sparsely populated terrain. The weakness of Libyan air defenses permitted relatively rapid degradation, reducing requirements for specialized electronic attack aircraft.[9] Future operational environments may lack these favorable characteristics. Conversely, Libya's operational strengths—for example, its air defenses' ability to leverage civilian networks to manage engagements—are likely to exist in the authoritarian areas where future NATO interventions are possible.[10]

Some vital "niche" operational capabilities simply do not exist in sufficient numbers within NATO. The United States provided fully 80 percent of refueling support during the course of Operation UNIFIED PROTECTOR, spurring France, Germany, and the Netherlands to announce cooperative tanker purchases.[11] Standoff precision-strike firepower was also lacking. France's SCALP (long range standoff cruise missile) naval cruise missile was not ready in time for Libya.[12] A report in *The Telegraph* suggested the UK expended a high proportion of its limited stock of Tomahawk Land Attack Missiles (TLAMs) in the first days of the conflict.[13] In contrast, by May 2011, two U.S. destroyers and one nuclear-powered Ohio-class submarine launched 199 TLAMs, ultimately launching 220 weapons in the course of the operation.[14] Similarly, the UK was short on advanced shorter-range munitions in some key categories.[15]

UNITED KINGDOM

The Royal Navy has dramatically declined in size by a third since 2000, but retains the desire and plans to remain a "balanced force" capable of naval airpower projection, limited amphibious operations, strategic nuclear deterrence, and sea control (see Figure 4-1). This goal remains even in view of the 2010 UK *Strategic Defence and Security Review* (SDSR) 8 percent de-

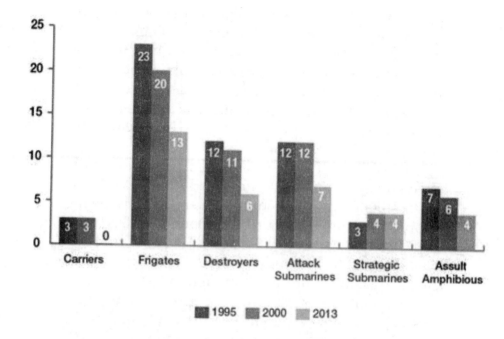

Sources: International Institute for Strategic Studies, "Chapter 4, Europe," *The Military Balance 2013*, Vol. 113, No. 1, pp. 89-198; International Institute for Strategic Studies, "NATO and Non-NATO Europe," *The Military Balance 2000*, Vol. 100, No. 1, pp. 35-108; and International Institute for Strategic Studies, "NATO," *The Military Balance 1995*, Vol. 95, No. 1, pp. 33-67.

Figure 4-1. UK (Total Ships by Category).

A key question, however, is whether a balanced force is ultimately in the strategic interests of the UK, or whether such a force should be abandoned in favor of a "cruising" navy requiring a greater number of frigates and destroyers and providing more naval presence in a greater number of places than the current fleet plan can accomplish. The costs associated with fielding two aircraft carriers and the air assets necessary to equip them, in addition to the costs of replacing the current fleet of ballistic missile submarines (SSBNs) with four new boats, will strain resources required for building surface combatants and attack submarines.[17] Considering the UK's global economic interests and its desire to remain closely aligned with the U.S. Navy, a force of less than 20 combatants might not suffice.

Upgrades to the Royal Navy will include fielding two new aircraft carriers carrying the F-35 Lightning II and the ongoing operation of the new, technologically advanced Type 45 destroyers.[18] Other upgrades include the continuing introduction of the five nuclear-powered, Astute Class attack submarines and the construction of the Type 26 Global Combat Ships.[19] Here as elsewhere in major NATO navies, numbers are being traded for capability.

When assessed against the roles articulated in the NATO *Alliance Maritime Strategy of 2011*—which includes deterrence and defense, crisis management, cooperative security, and maritime security—the Royal Navy presents a mixed story.[20] Continuing to move forward with both an aircraft carrier development program and a ballistic missile submarine program demonstrates national resolve to contribute to collective conventional and nuclear deterrence. However, the resources necessary to achieve these goals are to some degree harvested from savings gained from a significantly smaller escort and combatant fleet.

While the Type 45 destroyer is more capable than the Type 42s it replaces, there will be fewer of Type 45s, as there will be fewer Type 26 frigates to replace the Type 23s. This numerical decline creates presence deficits that impact the navy's ability to perform crucial traditional naval missions such as antisubmarine warfare (ASW) and antisurface warfare (ASUW), which underpin both conventional deterrence and cooperative and maritime security. Adding to a decline in traditional sea-control capabilities was the 2010 SDSR decision to eliminate the Nimrod maritime patrol aircraft from the inventory.

In summary, the Royal Navy continues to maintain a balanced fleet, one that looks strikingly like the U.S. Navy, except a fraction of its size. Its contributions on the high-end of the naval warfare operational spectrum (strategic deterrence, attack submarines, and anti-aircraft warfare [AAW] destroyers) are notable, while a declining number of surface combatants will bedevil its ability to remain globally postured and will contribute to naval missions of a more constabulary nature.

FRANCE

French defense policy in the post–Cold War era has tended toward greater equity among its armed services, what one analyst called the "gradual equalization" between French ground power and air and naval power.[21] Nevertheless, the overall downward trend in fleet size is clear (see Figure 4-2). In 2001, chief of staff of the French Navy Admiral Jean-Louis Battet identified a "2015 model" for the navy with a target fleet of 80 warships; the current trajectory is far more limited.[22]

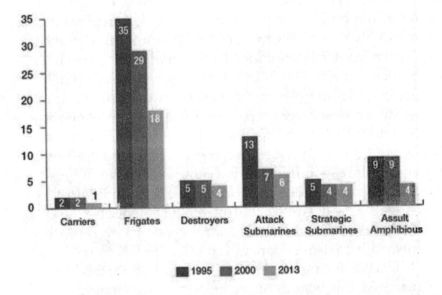

Sources: International Institute for Strategic Studies, "Chapter 4, Europe," *The Military Balance 2013*, Vol. 113, No. 1, pp. 89-198; International Institute for Strategic Studies, "NATO and Non-NATO Europe," *The Military Balance 2000*, Vol. 100, No. 1, pp. 35-108; and International Institute for Strategic Studies, "NATO," *The Military Balance 1995*, Vol. 95, No. 1, pp. 33-67.

Figure 4-2. France (Total Ships by Category).

Generally, the French Navy is currently faring better than land or air forces, but the declining share of French wealth spent on national defense—2.8 percent of gross domestic product (GDP) in 2008 and 1.76 percent in 2013—has inevitably impacted the fleet. And while the "main battery" of the French Navy—its aircraft carrier and 10 submarines—remain untouched, France's surface fleet will lose three destroyers and one amphibious ship. If there is any good news on this front, it is that France's 2008 defense white paper called for deeper cuts in fleet size and, unlike the

Royal Navy, the French Navy will not face, in the near term, the budgetary pressure of having to replace its relatively new SSBN force of four boats. Although the French fleet is shrinking, its international responsibilities remain. The 2013 white paper defined French geographic interests as "the European periphery, the Mediterranean area, a part of Africa—from Sahel to Equatorial Africa—the Persian Gulf and the Indian Ocean."[23] This perceived gap between strategic vision and actual capabilities has led some analysts to suggest that congruence between British and French interests, as well as a desire to control procurement costs and improve coalition interoperability, is driving France toward increasing cooperation with the UK.[24]

The 1998 Anglo-French Saint Malo declaration announced the beginning of heightened cooperation.[25] Attempts to establish effective cooperation on aircraft-carrier procurement and operations consumed much of the last decade. By 2007, an Anglo-French consortium looked to build three carriers for purchase by the two governments to maximize interoperability, but this plan did not come to fruition. Rumors that the two countries would actually share individual warships were again raised but quickly deflated in 2010.[26]

In contrast to the UK, which has primarily exported major warships and aircraft as second-hand articles to close British Commonwealth allies, France's defense industry competes actively to sell major platforms in the global market. The state-owned shipbuilder DCNS is set to deliver six Scorpène-class diesel-electric submarines to the Indian Navy starting in 2015.[27] The Indian order supplements two each already delivered to the Malaysian and Chilean navies.[28]

Additionally, France's DCNS shipbuilder and Italy's Fincantieri have been cooperating on the multimission frigate (FREMM) program. (At one point, France was planning to build 19 of these ships, but cuts in the ensuing years have dropped the buy to only 8.[29]) This industrial capacity augurs well for France, regardless of whether it increases the size of its navy, as international sales will protect a minimum level of shipbuilding capacity that is increasingly at risk in the UK.

With respect to NATO's stated maritime roles, the French Navy punches at a weight similar to the Royal Navy, though the French Navy's capacity for sea-control missions is somewhat better because of the numbers and age of its surface escort ships. Additionally, the French Navy's amphibious capabilities resident in its three Mistral-class LHDs and its one Foudre-class LPD provide a limited capacity for crisis response and humanitarian intervention. France's blue-water power-projection capability gives it the option of projecting power far from home waters, something the Royal Navy appears very much to desire as it proceeds to build its two Queen Elizabeth–class carriers.

Essentially, the Royal Navy and the French Navy are roughly equally sized and structured. Yet to many observers, the Royal Navy is in distress and the French Navy sails in relatively calmer waters. This stems at least in part from the pressure of history and the place of the Royal Navy in the hearts of average Englishmen.

GERMANY

Unlike the Royal Navy and French Navy, Germany lacks a history and culture (since World War II) of a "balanced" fleet capable of the full range of modern naval operations. With no carrier or amphibious fleet to speak of, and without a sea-based nuclear deterrent, the German navy historically has focused on sea-control missions centered around ASW, ASUW, and maritime security. While the number of ships devoted to these missions has fallen from 28 to 23 since 2000, the most precipitous decline has occurred within the submarine force, with older submarines having been replaced by four more-sophisticated submarines (Type 212As), and with two on order. (See Figure 4-3.)

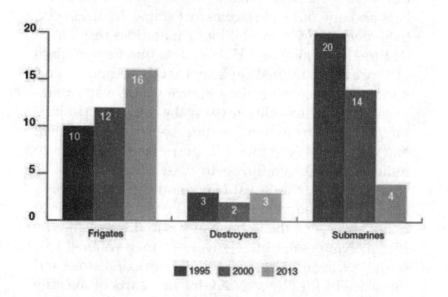

Sources: International Institute for Strategic Studies, "Chapter 4, Europe," *The Military Balance 2013*, Vol. 113, No. 1, pp. 89-198; International Institute for Strategic Studies, "NATO and Non-NATO Europe," *The Military Balance 2000*, Vol. 100, No. 1, pp. 35-108; and International Institute for Strategic Studies, "NATO," *The Military Balance 1995*, Vol. 95, No. 1, pp. 33-67.

Figure 4-3. Germany (Total Ships by Category).

Chief of Staff of the Navy Vice Admiral Axel Schimpf wrote in 2011 that Germany's armed forces in general, and the navy in particular, are favoring "width over depth" (or capability over capacity).[30] For the navy, which retained a greater share of its force structure than the other services as a result of recent budget cuts, this has meant continuing to build sophisticated air-independent propulsion diesel attack submarines for both domestic and international sale while maintaining a force of frigates and destroyers for blue-water operations focused mainly on ASUW and ASW. In fact, one reason the surface fleet appears to be maintained in the numbers it has been stems from the aggregate loss in ASUW power because of the smaller submarine force.

On the high-end of the operational spectrum, the three F124 Sachsen–class AAW destroyers are equipped with the Evolved SeaSparrow missile, an antiship defense missile, and the Standard Missile-3 Block IIA for point and area air defense. Of note, these ships integrate an active phased array radar with search and missile guidance capabilities, providing protection against both advanced aircraft and cruise missiles with reduced radar cross sections. When operating out of area, the German navy will likely deploy an F124 to provide air and missile defense to other less-capable German surface combatants. An interesting development in Germany has been the debate surrounding planning for the "common" procurement of a joint support ship (JSS). According to Vice Admiral Schimpf, such a ship (akin to a U.S. LPD) would have several missions, including military evacuation operations, humanitarian aid from the sea, conduct of land operations from the sea, special forces employments, and "ensured military maritime deployability."[31] Currently two are planned, but they have not been funded because of debate over the cost to be allotted to Germany's army and air force.

The German navy's contributions to NATO's maritime roles fall mainly within the lower end of the operational spectrum. Germany's cruising navy provides little in the way of power projection, but, for out-of-area operations, the fleet adds to alliance maritime security and cooperative security, and, though the sea-control capabilities resident in these platforms, it can contribute to collective defense. Should Germany proceed with the JSS, it would have greater capacity to engage in maritime humanitarian assistance operations and to marginally increase its ability to project power.

The German navy — unlike the Royal and French navies — does not have a desire to be a balanced force capable of significant power projection, amphibious operations, and strategic deterrence. As its aims have been historically more modest, they have been more capable of being supported. To the extent that Germany continues to support NATO maritime operations of a largely constabulary nature, Germany's contributions to NATO remain consistent. The interesting question is not whether the navy supports Germany's worldview and view of itself; it is whether a nation as powerful, rich, and networked as Germany is underinvesting in naval power while free-riding on the backs of U.S., UK, and French naval capabilities to a greater extent than other European nations.

SPAIN

In the last decade, Spain appeared to be a nation putting its best defense (and naval) foot forward.

With a moderately rising defense budget in the first half of the decade and a number of international ship-building partnerships underway, the Spanish navy was quantitatively and qualitatively improving. This progress was halted by the global economic crisis that has caused Spain to cut defense spending three times since 2008: by 3 percent in 2009, by 6.2 percent in 2010, and by nearly 17.6 percent in 2012.[32] Interestingly, Spain has not announced any plan to reduce commit-ments, missions, or capabilities, deciding instead to go the route of other European nations, which is to favor cuts in capacity rather than capability.[33]

The financial crisis–induced cuts were made to a budget that was already one of the worst within NATO in terms of meeting the 2 percent-of-GDP de-fense-spending goal agreed to by NATO members in 2002. In 2010, Spain spent just 0.72 percent of its GDP on defense, with no year in the previous 5 years even coming close to approaching 1 percent.[34]

Spain has sought a balanced navy, operating a flagship aircraft carrier (*Príncipe de Asturias*), five AE-GIS-enabled guided missile destroyers (DDGs) of the Álvaro de Bazan class, six frigates of the Santa Ma-ria class—a Spanish version of the U.S. Navy's FFG-7-class guided missile frigates—and four Galerna-class diesel submarines, in addition to three principal amphibious ships (see Figure 4-4).[35]

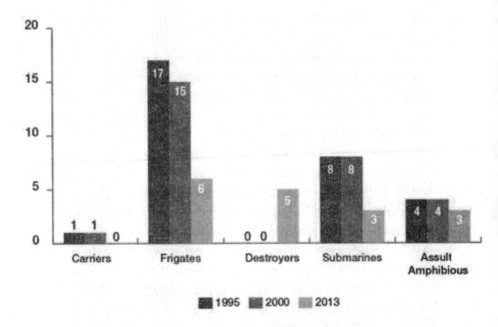

Sources: International Institute for Strategic Studies, "Chapter 4, Europe," *The Military Balance 2013*, Vol. 113, No. 1, pp. 89-198; International Institute for Strategic Studies, "NATO and Non-NATO Europe," *The Military Balance 2000*, Vol. 100, No. 1, pp. 35-108; and International Institute for Strategic Studies, "NATO," *The Military Balance 1995*, Vol. 95, No. 1, pp. 33-67.

Figure 4-4. Spain (Total Ships by Category).

Spain's shipbuilding industry has competed strongly on the world market, cooperating with France's state-owned DCNS on the Scorpène submarine program, which morphed into Spain's S-80 class, four of which remain under construction even in light of ongoing defense cuts.[36] Additionally, Spanish shipbuilders are constructing the second of two 27,000-ton Canberra-class LHDs for the Royal Australian Navy.[37]

The primary threat to Spain's navy from ongoing budget woes is its inability to modernize and maintain fleet size. Insufficient funds in 2012 caused the navy to cannibalize one of its four Galerna-class submarines for parts to keep the other three boats operational.[38] Additionally, five vessels were decommissioned in 2012, and in early-2013, even the *Príncipe de Asturias* was decommissioned. The 2012 budget virtually eliminated spending for the majority of Spain's 19 major defense-wide procurement programs.[39]

Spain's contributions to NATO's maritime roles, while not in the class of the UK or France, remain relatively strong in what is admittedly an increasingly weak field. The loss of its aircraft carrier and the decline in ship numbers essential to complex ASW and ASUW missions have been somewhat offset by the emergence of the five highly capable F100 destroyers equipped with the U.S. AEGIS system featuring the SPY-1D radar. Additionally, Spain's modest amphibious capability contributes to both power projection and humanitarian missions.

ITALY

Italy historically fields a balanced fleet with aircraft carriers, diesel submarines, surface combatants, and amphibious ships. Without an undersea strategic deterrent, its navy resembles that of Spain, though somewhat larger and more powerful. Like the other navies surveyed, it is getting smaller. Its shrinking predates the global financial crisis, but financial restraints have clearly accelerated the condition.

The Italian navy has a goal of allocating 50 percent of its budget to personnel costs; 25 percent to investment and procurement; and 25 percent to operations, maintenance, and training. However, personnel costs have consumed upward of 70 percent of the budget in recent years, even as the navy strove to keep important acquisition programs going. This has inevitably squeezed the operations, maintenance, and training budget, which was allotted only 11.2 percent of the 2012 budget.[40]

In May 2012, in testimony before parliament, the navy's chief of staff called the current force structure "unsustainable," announcing plans to retire 26 to 28 ships by 2017.[41] Recent austerity measures have seen major purchases reduced or delayed. The head of Italy's navy stated that "funding issues" exist with the final two of the six frigates Italy has thus far ordered from the Franco-Italian FREMM program.[42] Two more German U212A submarines will be purchased, but likely at the cost of retiring the Sauro-class boats, reducing the current submarine fleet from six to four.[43] Italy initially planned to purchase six Horizon-class AAW destroyers produced by an earlier joint venture with France, but by 2006 judged two sufficient for escort of its carriers or amphibious warships.[44]

With respect to NATO maritime roles, Italy, like the UK, has favored power projection over sea control. This is plain from Italy's current order of battle, which features an aging and shrinking frigate force (see Figure 4-5). The FREMM program appears designed to bring additional balance to the fleet by increasing sea-control capabilities. The navy's chief of staff has reiterated the service's strong desire for 10 FREMM ships, while admitting that Italy's shaky finances threaten this goal.[45]

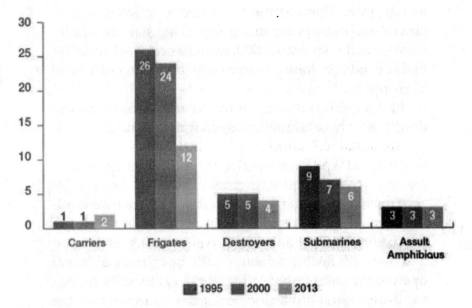

Sources: International Institute for Strategic Studies, "Chapter 4, Europe," *The Military Balance 2013*, Vol. 113, No. 1, pp. 89-198; International Institute for Strategic Studies, "NATO and Non-NATO Europe," *The Military Balance 2000*, Vol. 100, No. 1, pp. 35-108; and International Institute for Strategic Studies, "NATO," *The Military Balance 1995*, Vol. 95, No. 1, pp. 33-67.

Figure 4-5. Italy (Total Ships by Category).

With respect to higher-end missions including air and missile defense, the new Andrea Doria–class destroyers are a formidable escort with capabilities tested against advanced cruise-missile targets. However, they number only two. When the Durand de la Penne class retires its medium-range surface-to-air missiles in "5 to 6 years," Italy will be left with only two effective anti-air escorts.[46]

The Italian navy is headed in the same direction as the UK, France, and Spain: it will have a technologically advanced naval force structure that is balanced among power projection, amphibious operations, and classic sea-control missions but that is dramatically smaller than its year 2000 predecessor.[47] Like other NATO navies, Italy believes the prudent path is to keep the basic architecture for a fleet with global influence while procuring ships in numbers that raise doubt as to how influential such a navy could be.

WHERE STANDS NATO?

The major navies of the NATO alliance (including the U.S. Navy) have much in common. With the exception of Germany, the focus remains on having a "balanced fleet" capable of the spectrum of naval operations from cooperative security through war at sea and power projection. Of course, France and the UK continue to maintain a strategic nuclear deterrent through ballistic missile submarines.

It is not inconceivable that in the near future (early-2020s), only the United States, France, and the UK will routinely operate aircraft carriers within NATO, with the UK program seemingly always on the edge of the budgetary chopping block. The difficulty NATO had in waging air surveillance and strike from the sea during the Libya operation, without a U.S. carrier, is likely to be exacerbated. But even if the UK and France continue to operate carriers, the likely cost will be reduced global presence in maritime security and constabulary missions that require a larger fleet of blue-water surface combatants. Those countries are likely to be willing to pay that price, as the ability to contribute carrier-striking power to U.S.-led operations—NATO and otherwise—continues to provide a *sine qua non* of naval relevance.

The desire to maintain a balanced fleet—irrespective of its size—cannot help but raise the question of whether what is driving these decisions is as much about national pride as national or alliance strategy. Certainly, eliminating either their aircraft carriers or their ballistic missile submarines would free up funds for an expanded French or British fleet of surface combatants.

Moreover, China's naval renaissance impacts NATO nations' force-structure decisions. As the United States turns more of its interest to the Pacific, baseline security requirements in the Mediterranean will become more important to Europe's NATO navies, perhaps creating greater incentive to resource them. Additionally, both France and the UK see themselves as global nations with global interests that extend far into the Pacific. If these nations perceive China's rise as threatening these interests, they will likely find their navies too small to provide any real impact, given the great distances involved and the paucity of ships to maintain constant presence. There is a real tension between global presence and a "balanced fleet," one that currently only the United States is able to resolve, and barely at that.

The United States must come to grips with the likelihood that, even with its Navy declining in size, over time, it will comprise an increasing percentage of alliance striking power. The 2007 maritime strategy designated the Indian Ocean and Persian Gulf as well as the Western Pacific as the U.S. Navy's two major operational hubs, recognizing in print what had been practiced operationally since the First Gulf War. This posture leaves the Mediterranean routinely without carrier or amphibious striking power, something that was evident in the early days of the Libya campaign.

With European carrier-striking power likely to wane, the United States will find itself trying to stretch its 11-carrier fleet across three operational hubs, something it did in the 1980s with 15 carriers. While 11 aircraft carriers are currently written into public law as the minimum number the Navy must maintain, Congress can even change that if it sees fit.

Absent a crisis or a threat that manifests itself in large part as a naval threat, Europe is unlikely to return to large, balanced fleets. Once lost, however, it could take decades to rebuild naval force structure because of the capital-intensive nature of shipbuilding and the time it takes to build sophisticated, modern warships in an increasingly small number of capable shipyards. NATO members should be wary about continuing declines in force structure. While current efforts to coordinate militaries ("pooling and sharing") may on the surface seem beneficial, care must be taken that such efforts are not simply window dressing for further decline.

ENDNOTES - CHAPTER 4

1. This essay was originally published on September 18, 2013.

2. Compare this number with two authoritative ship counts bracketing the Falkland Islands conflict, in which the Royal Navy fielded 67 combatants in 1980 and 56 in 1985. See Gavin Berman, *Defence Statistics 2000: Research Paper 00/99*, London, UK: House of Commons Library, December 21, 2000, available from *www.parliament.uk/briefing-papers/RP00-99*.

3. Jorge Benitez, *National Comparison of NATO Strike Sorties in Libya*, NATO Source, Washington, DC: Atlantic Council of the United States, August 22, 2012, available from *www.acus.org/natosource/national-composition-nato-strike-sorties-libya*; and "France to Withdraw Aircraft Carrier From Libya Ops," Agence France-

Presse, August 4, 2011, available from *www.defensenews.com/article/20110804/DEFSECT03/108040307/France-Withdraw-Aircraft-Carrier-From-Libya-Ops>*.

4. For quantitative analysis of maintenance cycles and alternatives to the U.S. Nimitz class, see Roland J. Yardley *et al.*, *Increasing Aircraft Carrier Forward Presence: Changing the Length of the Maintenance Cycle*, Santa Monica, CA: RAND Corporation, 2008, available from *www.rand.org/content/dam/rand/pubs/monographs/2008/RAND_MG706.pdf*.

5. The STOVL carriers remaining were Britain's LPD, H.M.S. *Albion*, and the landing platform helicopter H.M.S. *Ocean*, alongside the French LPD *Tonnere*, the U.S. Wasp-class LHDs *Kearsage* and *Bataan*, and the San Antonio-class *Mesa Verde*. See "Italy Removes Aircraft Carrier from Libya Campaign," Agence France-Presse, July 7, 2011, available from *www.defensenews.com/article/20110707/DEFSECT05/107070311/Italy-Removes-Aircraft-Carrier-from-Libya-Campaign*.

6. The *Mesa Verde* and the *Bataan* had already deployed in March. See William H. McMichael, "Bataan ARG Heads to Libya Duty in Med," *Marine Corps Times*, March 23, 2011, available from *www.marinecorpstimes.com/news/2011/03/navy-libya-bataan-arg-deploys-early-032311w/*. By the time it returned from Libya in May 2011, *Kearsarge* had been deployed for 9 months and continuously at sea for 4 months in support of Operation UNIFIED PROTECTOR. See William H. McMichael, "Kearsarge Back from Extended Deployment," *Navy Times*, May 16, 2011, available from *www.navytimes.com/news/2011/05/navy-kearsarge-ready-group-home-051611w/*.

7. The *Albion* (lead ship of its class) is 176 meters in length, while the *Tonnere* is 199 meters long. Both are around 21,000 tons and both are appreciably smaller than the 40,000-ton, 250-meter *Kearsarge* and *Bataan*. If deployment of STOVL aircraft is possible at all from these smaller ships, it would require substantial modifications.

8. For the regime's pre–Civil War forces, see International Institute for Strategic Studies, "Chapter Five: Middle East and North Africa," *The Military Balance*, Vol. 110, No. 1, pp. 262–263.

9. Craig Hoyle, "NEWS FOCUS: How 'Odyssey Dawn' Tamed Libya's Air Defences," *Flightglobal*, March 28, 2011, available from *www.flightglobal.com/news/articles/news-focus-how-odyssey-dawn-tamed-libyas-air-defences-354736/*.

10. See Jeff Kassebaum, "The Art of SEAD: Lessons from Libya," *Journal of Electronic Defense*, Vol. 34, No. 12, December 2011, pp. 58–62.

11. "NATO Trio Team Up to Boost Air Refueling Capacity," Agence France-Presse, April 18, 2012, available from *www.defensenews.com/article/20120418/DEFREG01/304180013/NATO-Trio-Team-Up-Boost-Air-Refueling-Capacity*.

12. The program was previously known as Missile de Croisière Naval. See Richard Scott, "MBDA Completes First Scalp Naval Test-Firing," *Jane's International Defence Review*, June 17, 2010.

13. Thomas Harding, "Libya: Navy Running Short of Tomahawk Missiles," *The Telegraph*, March 23, 2011, available from *www.telegraph.co.uk/news/worldnews/africaandindianocean/libya/8400079/Libya-Navy-running-short-of-Tomahawk-missiles.html*. Harding's figure of 12 missiles is corroborated in Jeremiah Gertler, *Operation Odyssey Dawn, Libya: Background and Issues for Congress*, Washington, DC: Congressional Research Service, March 30, 2011, p. 18, available from *www.fas.org/sgp/crs/natsec/R41725.pdf*.

14. Norman Polmar, "The Latest Conflict," *Proceedings of the U.S. Naval Institute*, Vol. 137, No. 5, May 2011, pp. 166–167; and "Navy Replenishing Tomahawk Stockpile Used in Libya," *Defense Daily International*, June 14, 2012.

15. British analysts felt the Royal Air Force's supply of dual-mode seeker-equipped Brimstone and Paveway IV air-to-ground weapons needed restocking during or shortly after the end of operations. See *Lessons Offered From the Libya Air Campaign*, London, UK: Royal Aeronautical Society, 2012, p. 4, available from *aerosociety.com/Assets/Docs/Publications/SpecialistPapers/LibyaSpecialistPaperFinal.pdf*; and "UK to Buy at Least 500 Paveway IVs To Restock after Libya," *Jane's Defence Weekly*, September 15, 2011. (Although it appears that the UK's supply of air-launched Storm Shadow cruise missiles appears to have been sufficient.)

16. Nicholas Watt, "Next Generation of Nimrod 'Spy In the Sky' Surveillance Planes To Be Scrapped," *The Guardian*, October 17, 2010, available from *www.guardian.co.uk/uk/2010/oct/17/next-generation-nimrod-scrapped*.

17. In response to questions about the costs associated with the SSBN programs, British Defense Secretary Philip Hammond insists that "The government remains 100% committed to maintaining and renewing the Trident system." See "U.S. Defense Chief Bashes Idea of Reducing SSBN Fleet," *Global Security Newswire*, July 15, 2013, available from *www.nationaljournal.com/global-security-newswire/u-k-defense-chief-bashes-idea-of-reducing-ssbn-fleet-20130715*.

18. The Type 45 is built primarily as an AAW combatant capable of local and area fleet defense. Capable of controlling fighter aircraft, it can coordinate fleet AAW operations and should be considered roughly comparable to a U.S.-guided missile destroyer. It is equipped with long-range weapon systems to intercept increasingly sophisticated and maneuverable missiles. The Type 45 destroyer will be able to operate an embarked helicopter.

19. Due to begin joining the fleet in 2021, the Type 26 frigates will completely replace the Type 23 frigates. The Type 26 is planned in three variants: an ASW variant, an AAW variant, and a general purpose variant.

20. NATO, "Alliance Maritime Strategy," March 18, 2011, available from *www.nato.int/cps/en/natolive/official_texts_75615.htm*.

21. Lutz Unterseher, *Europe's Armed Forces at the Millennium: A Case Study of Change in France, the United Kingdom, and Germany*, Washington, DC: Project on Defense Alternatives, 1999.

22. Jean-Louis Battet, "The French Navy in the Twenty-First Century," *NATO's Nations and Partners for Peace*, No. 3, 2001, pp. 23–29.

23. Government of France, *Livre Blanc: Défense et Sécurité Nationale 2013* (White Paper: Defense and National Security-2013), Paris, France, 2013, p. 82, available from *www.livreblancdefenseetsecurite.gouv.fr/pdf/le_livre_blanc_de_la_defense_2013.pdf*.

24. Ben Jones, *Franco-British Military Cooperation: A New Engine for European Defence?* Paris, France: European Union Institute for Security Studies, February 2011, pp. 13–17.

25. "Anglo-French Military Pact," BBC News, December 4, 1998, available from *news.bbc.co.uk/2/hi/uk_news/politics/227598.stm.*

26. Tim Fish, "Anglo-French Carrier Co-operation Moves Forward," *Jane's Navy International,* November 16, 2007; and Dave Clark, "France, Britain: No Plans to Share Aircraft Carriers," Agence France-Presse, September 3, 2010, available from *www.defensenews.com/apps/pbcs.dll/article?AID=20109030305.*

27. The project has not been without political trouble. See Vivek Raghuvanshi, "Indian Minister Rejects Charges Connected with Sub Deal," *Defense News,* July 19, 2012, available from *www.defensenews.com/apps/pbcs.dll/article?AID=2012306190001.*

28. "SSK Scorpene Class Attack Submarine, France," *Naval Technology,* available from *www.naval-technology.com/projects/scorpene/*; and "DCNS, Navantia Part Ways on Submarines," *United Press International,* November 15, 2010, available from *www.upi.com/Business_News /Security-Industry/2010/11/15/DCNS-Navantia-part-ways-on-submarines/UPI-29901289850253/.*

29. Jean-Dominique Merchet, "Marine : Ce Sera 8 Fremm au Lieu de 11: Le Nombre de Nouvelles Frégates Multimissions Sera Réduit" ("Marine: It Will Be 8 FREMM Instead of 11: The Number of New Multimission Frigates Will Be Reduced"), *Secret Defense,* available from *www.marianne.net/blogsecretdefense/Marine-ce-sera-8-Fremm-au-lieu-de-11_a1028.html.*

30. Axel Schimpf, "The German Navy of the Future," *European Security and Defence 3–4, 2011,* pp. 7–12, available from *www.scribd.com/doc/144757062/European-Security-Defence-Magazine-Issue-Nos-3-4-2011#scribd.*

31. Schimpf, "The German Navy of the Future," p. 12.

32. International Institute for Strategic Studies, "Chapter Four: Europe," *The Military Balance 2011,* Vol. 111, No. 1, pp. 144–

147. See also, "Chapter Four: Europe," *The Military Balance 2013*, Vol. 113, No. 1, p. 95.

33. Caroline Baxter *et al.*, *NATO and the Challenges of Austerity*, Santa Monica, CA: RAND Corporation, 2012, p. 46.

34. *Ibid.*, p. 51.

35. The Príncipe de Asturias was taken out of service in early 2013. See "El Príncipe de Asturias' llega a Ferrol Para Su Desarme" ("The Príncipe de Asturias Arrives in Ferrol for Its Disarmament"), El País, Spain, February 8, 2013, available from *ccaa. elpais.com/ccaa/2013/02/08/galicia/1360342099_511362.html.*

36. Christina Mackenzie, "Spanish S-80 Subs Sailing Forward," *Aeropspace Daily and Defense Report*, July 18, 2012. The S-80 class has had challenges in construction, including the first hull displacing more than 100 tons than was designed, which could impact the boat's ability to submerge and resurface. Additionally, there have been reports of problems in the boat's air-independent propulsion system. See "GD To Help Fix Spanish Navy's Overweight Issue of S-80 Submarine" *Naval Techology*, June 7, 2013, available from *www.naval-technology.com/news/newsgd-to-help-fix-spanish-navy-overweight-issue-s80-submarine.*

37. David Ing and Richard Scott, "Arrested Development: Austerity Stunts Spain's Naval Industrial Sector," *Jane's Navy International*, September 7, 2012.

38. "Durante este tiempo, las necesidades de mantenimiento de los buques han hecho que se empleen piezas del 'Siroco' en el resto de la flota de los S-70. Defensa admitió en su día que se iba a utilizar piezas del 'Siroco' para las carenas del 'Tramontana' y 'Galerna'" ("During this time, the maintenance needs of the ships have required that parts from the 'Siroco' be used in the rest of the S-70 fleet. The ministry of defense admitted that it was going to use the parts for the hulls of the 'Tramontana' and 'Galerna'.") See Jose Alberto Gonzalez, "La Armada da de Baja el 'Siroco' y Centra Sus Esfuerzos en Los Submarinos S-80" ("The Navy Withdraws the 'Sirocco' and Focuses Its Efforts on the Submarines S-80"), La Verdad, August 5, 2012, available from *www. laverdad.es/murcia/v/20120508/cartagena/armada-baja-siroco-centra-20120508.html.*

39. Pedro Arguelles, Comparencencia Presupuestos 2012 (2012 Budgets Appearance), testimony before the Spanish Congress of Deputies, April 17, 2012, available from *rojoygualda.files. wordpress.com/2012/04/secdef170412.pdf.*

40. Schmitt, "Italian Hard Power: Ambitions and Fiscal Realities," p. 8.

41. Luca Peruzzi, "Cuts Loom for Italian Navy," *Jane's Navy International,* June 11, 2012.

42. Luca Perruzzi, "Fincantieri Launches First Italy's FREMM Multi-Mission Frigate," European Security and Defense Press Association, July 23, 2011, available from *www.esdpa.org/2011/07/ fincantieri-launches-first-italy%E2%80%99s-fremm-multi-mission- frigate/.*

43. Luca Perruzzi, "Interview: Admiral Luigi Binelli Mantelli, Chief of the Italian Navy," *Jane's International Defense Review,* March 16, 2012.

44. "DCN Launches Final Horizon," *Jane's Navy International,* July 13, 2006.

45. Perruzzi, "Interview: Admiral Luigi Binelli Mantelli."

46. *Ibid.*

47. One example of this desire to hold onto the trappings of a high-end navy is the recent decision by the Italian government to reduce its purchase of F-35Bs for the Italian Navy to 15 after the head of Italian naval aviation claimed that the minimum number of STOVL aircraft required was 22. See Tom Kington, "Italian Navy, AF Head for F-35B Showdown," *Defense News,* May 15, 2012, available from *www.defensenews.com/article/20120515/ DEFREG01/305150010/Italian-AF-Navy-Head-F-35B-Showdown.*

CHAPTER 5

GERMAN HARD POWER: IS THERE A THERE THERE?[1]

Patrick Keller

The opinions expressed in this chapter should be attributed to the author alone. The author thanks his research assistant, Aylin Matlé, for her support.

KEY POINTS

- German ambivalence on the use of military power continues to bedevil German politicians and leaders.
- A stagnant defense budget will be a challenge to the German defense ministry's plan to establish a leaner, more flexible, and more deployable German armed forces.
- As Europe's economic leader and central political actor, Germany should guide the way in reversing the problematic decline in European hard power.

Two very different stories are in competition for the "grand narrative" of current German security policy. The first could be called "look how far we've come" and goes like this: Since reunification restored the state to full sovereignty in 1990, a thriving Germany has accepted its increasing share of responsibility in international security affairs. It has done so gradually — mindful of its historic baggage — but efficiently. After the 1994 breakthrough decision by the Federal Constitutional Court to allow out-of-area deployments of the Bundeswehr (German armed forces), the forces have been partaking in many North Atlantic Treaty Organization (NATO) and European Union (EU) missions, including the wars in Kosovo and Afghanistan and the fight against piracy off the coast of Somalia.[2] Currently, Germany deploys about 6,200 troops in missions abroad; it is the third largest contributor to the International Security Assistance Force (ISAF) in Afghanistan and the lead nation in the NATO-led Kosovo Force (KFOR). Thus, contemporary Germany has finally established itself as a "normal nation" that contributes to international stability. It does so — if necessary — by military means as well, and certainly in a manner that is commensurate with its size and economic strength.

The other story could be called "too little, too late" and scoffs at these alleged achievements. From this perspective, German security policy during the last 25 years has always oscillated between two conflicting conclusions drawn from German history. One is never again to stand opposed to the United States and Germany's (major) European neighbors; the other is never again to experience war. Hence, although Germany has made military contributions to international missions, it has never done so by its own initiative. Germany's allies (mostly the United States) and partners in the EU had to drag Germany into its commitments. As a consequence, German leaders of various political persuasions have always tried to commit as few troops with as many caveats (such as restricted rules of military engagement) as possible without losing face among allies and friends. One can debate whether this is a prudent strategy and whether it worked well, but few would argue that it is a policy befitting the most prosperous, populous, and politically influential nation-state in the EU.

Every German security policy expert puts forward a version of one of these two stories or a combination of both, depending on circumstance. (The politically savviest tell the first story to international audiences, while saving the second story for domestic consumption.) This unresolved "grand narrative" debate betrays German policymakers' fundamental insecurity about their country's role in the world and about the proper bearing for a leading power. What is even more curious, however, is how abstract this debate really is: very few talk seriously about the fundamentals of German security and defense policy — that is, about Germany's military capabilities.

Both narratives implicitly assume that German military capabilities exist in sufficient number and quality to give policymakers a broad range of strategic choices, while in fact such hard power assets are waning in Germany and almost everywhere else in the West.[3] If current trends continue, a different pair of competing stories might occur because "we cannot, as we are simply lacking the capabilities to do so," versus the more sophisticated, "We cannot fight anymore because we do not want to and took all necessary steps to prevent us from having those capabilities." Either way, the continuation of current trends will result in calamity — not just for German security interests but also for the overall stability of a liberal international system.

GERMAN ARMED FORCES IN TIMES OF AUSTERITY

Since 1990, the Bundeswehr has been undergoing constant reform. Main drivers of these reforms were the incorporation of the East German army (German

Democratic Republic's National People's Army) into the Bundeswehr, the adaptation to new tasks in a changed security landscape after the Cold War, and the constraints of a limited defense budget. In fact, the military and the German ministry got so tired of the unending reform cycles that current minister of defense Thomas de Maizière prefers instead to call his reform a new orientation (*Neuausrichtung*). Tellingly, this latest wave of Bundeswehr reform did not originate with a security-political decision by the defense minister but with a budget decision by the finance minister.

This daunting requirement propelled then–minister of defense Karl-Theodor zu Guttenberg to initiate the most far-reaching reform of the Bundeswehr since its founding in 1955. In a first step, he killed one of his conservative party's sacred cows: conscription. The practicality (and feasibility) of maintaining a conscription army in a post–Cold War security environment that required leaner and more professional forces had been contested for years. Sold as a cost-saving exercise in dramatic financial times by Germany's most popular minister, protests against the change were suddenly soft. As it turned out, however, ending conscription did not save money but created extra cost for recruiting and maintaining salary levels competitive with the private sector.

In response to the 2008 global financial crisis and the ensuing European debt crisis, German Chancellor Angela Merkel's government adopted a constitutional amendment limiting new federal debt to 3.5 percent of gross domestic product (GDP). To comply with this break on debt (*Schuldenbremse*), in 2010, Finance Minister Wolfgang Schäuble prescribed every ministry an exact amount of money to be saved over the following

4 years. In relation to its overall budget, defense had to cut the most: €8.3 billion until 2014. Considering that the annual German defense budget is only about €30 billion, the prescribed reduction was substantial—especially for a military establishment already existing on limited means.

Thus, other elements of zu Guttenberg's reform package—downsizing the armed forces, reducing procurement of new weapons systems and platforms, trimming resources for research and development (R&D) for future systems, and increasing cooperation with EU partners on military matters ("pooling and sharing")—became even more relevant. The actual concepts behind those general ideas, however, remained nebulous. When zu Guttenberg had to resign in March 2011 over allegations of plagiarism in his dissertation, it fell to de Maizière, zu Guttenberg's successor, to develop a strategy that satisfied both the treasurers and the generals. Such a strategy, de Maizière decided, should consider Germany's negative demographic trend, should be derived from an analysis of Germany's political and security situation, and should be financially sustainable.

Surprisingly, de Maizière—who is one of Angela Merkel's closest advisers and was, in her first term, the chief of her chancellery—proved capable of working under less harsh conditions than assumed: the prescribed cuts of €8.3 billion were taken off the table. To the contrary, the administration and parliament even agreed to a slight increase in defense spending and to project more modest reductions over the next 2 years. (See Table 5-1.)

Year	Defense Spending (in billions of euros)	% of GDP
2006	27.87	1.2
2007	28.38	1.2
2008	29.45	1.2
2009	31.18	1.3
2010	31.11	1.3
2011	31.55	1.2
2012	31.70	1.2
2013	33.30	To be determined
2014	30.90 (projected)	To be determined
2015	30.40 (projected)	To be determined

Source: Federal Statistical Office of Germany, "Volkswirtschaftliche Gesamtrechnungen" (National Accounts), available from *www.destatis.de/DE/ZahlenFakten/Indikatoren/Konjunkturindika toren/VolkswirtschaftlicheGesamtrechnungen/vgr111.html*; Federal Ministry of Defence, "2012 Defence Budget," August 14, 2012, available from *www.bmvg.de/portal/a/bmvg/!ut/p/c4/NYxBCoN-ADAB_tJFCL976iqqXJdYQg0kqa3b7ffVQhjkNDExw4diEMeTrqD-DA-JF-_qXZGicT lyOoSLXUqATJIlydjxXrpUZ-dN0zLVQyblFJlfl_wfueK8NIDrv Z6wSyiBio/*; and German Bundestag, *Unterrichtung durch die Bundesregierung* (Briefing by the Federal Government: The Federal Budget 2011 to 2015), December 8, 2011, available from *dipbt.bundestag.de/dip21/btd/17/066/1706601.pdf*.

Table 5-1. German Defense Spending as a Percentage of GDP.

From 1991 until 1997, German defense spending was continually decreasing (from about €28 billion to €23 billion and, correspondingly, from approximately 2 percent of GDP to 1.6 percent). With the Kosovo War, the "peace dividend" era was over. Since 2001, defense spending has been on a slow but steady rise, with only minor cuts in 2003 and 2010. The financial crisis, starting in 2008, did not have a discernible effect on this trend. Indeed, the projected cuts for 2014 and 2015 might yet be reversed — after all, the administration's original projected defense budget for 2013 was €31.4 billion, well below the €33.3 billion that was actually allocated.

At the same time, German increases in defense spending have remained modest and have not even offset the effects of inflation over the past 20 years. In real terms, defense spending has been decreasing. Moreover, with defense spending at around 1.25 percent of GDP, Germany obviously does not make defense a budget priority. (The budget of the ministry of labor and welfare is more than four times the size of that of the ministry of defense.) Needless to say, Germany does not meet the pledge made by the NATO allies at the 2002 Prague NATO Summit to spend at least 2 percent of national GDP on defense.

The budget increase of about 5 percent in 2013 seems striking, but it is because of a significant rise in the personnel cost of federal employees and a projected rent hike for some buildings used by the armed forces. It is not a gain in substance for military planners;[4] in fact, the budget share allocated to the investment in actual defense-related capabilities (including not only military procurements but also R&D) has declined in both absolute and relative terms. In 2012, R&D and procurement constituted approximately 23.1 percent (€7.4 billion) of the total defense budget, but was reduced to 21.4 percent (€7.1 billion) for 2013. The figures for military procurements alone also reflect this, with a reduction from 17.2 percent to 15.4 percent (or €5.5 billion to €5.1 billion in absolute figures).[5]

In an effort to ease the budgetary squeeze, de Maizière proceeded to trim ministry structures and to downsize the armed forces. Upon completion of his new orientation in 2017, the Bundeswehr is envisioned to consist of no more than 185,000 active duty military and 55,000 civilian employees (down from 250,000 and 75,000, respectively, in 2010), with 10,000 soldiers deployable simultaneously in two areas of operation (up from 7,000).[6] In the new personnel structure, the army, air force, and navy will consist of approximately 62,000 soldiers, 32,000 airmen, and 16,000 sailors, respectively, and the Joint Support Service and the medical service will consist of roughly 46,000 and 19,000, respectively.[7] (The remaining members are distributed among equipment, infrastructure, human resources, and other services.)

This development is accompanied by reductions in military materiel through cuts in prospective procurement and decommissioning of active systems. Although the German Navy is to remain more or less the same (albeit at a lower level of personnel), these reductions will strongly affect Germany's army and air force.[8] (Table 2 shows some of the prospective changes.) To assess what this means for German defense policy, one needs to consider the strategic context of these changes.

System	Current or orginally planned number	New ceiling
Combat tank Leopard 2	350	225
Armored personnel carriers Puma/ Marder	410/70	350/0
Armored howitzer 2000	148	89
Multipurpose helicopter NH-90	122	80
Support helicopter Tiger	80	40
Eurofighter Thyphoon	177	140
Combat aircraft Tornado	185	85
Transport aircraft C-160/A400M	80/60	60/40
Multipurpose warship (MKS 180)	8	6
Naval mine countermeasures unit	20	10

Source: Federal Ministry of Defense, "Ressortbericht zum Stand der Neuausrichtung der Bundeswehr" (Interagency Report on the State of the Reorientation of the Bundeswehr), Bundesministerium der Verteidigung, May 8, 2013, p. 24, available from *www. bmvg.de/portal/a/bmvg/!ut/p/c4/NYvBCsIwEET_aDcBRerNEhSv-vdh4S9sQVpqkrJt68eNNDs7AO8 xj8Im1ye0UnFBObsUR7Uzn6QN-T3AO8cuG6QqREb_FMJeKjfRYPc05eGsUnocrATjLDllnWZgpzNU-ALWqVNr7T6R3-70-1qzUEfzb0fc Ivx8gOBJaR2/.*

Table 5-2. Change in German Defense Procurement.

STRATEGIC BACKDROP AND LEVEL OF AMBITION

According to Minister de Maizière, the cuts described in Table 5-2 are not primarily dictated by budget constraints but reflect security-political considerations. Using Germany's 2006 white book as a starting point, the minister outlined the strategic thinking that was to guide the "new orientation" in a series of documents and speeches. The most important of those are the Defense Policy Guidelines (DPG) and and the principles (*Eckpunkte*) papers, both published in May 2011.[9] They provide a rationale for the German military in the early-21st century by explaining Germany's vested interest in a stable liberal international order and by analyzing current and likely future threats to that order.[10] The ministry emphasizes that neither retrenchment nor the sole focus on traditional concepts of territorial defense are promising strategies in dealing with these challenges. Hence, Germany should take on a greater share of the burden in upholding global order, including military contributions to UN, EU, or NATO missions.

Consequently, the "new orientation" seeks to develop a sleeker force that is highly deployable and effective in crisis management and crisis resolution missions. "The ability to fight . . . is thus a benchmark for operational readiness," states the DPG.[11] Because of Germany's size and geostrategic position, the ability to fight cannot be limited to a few specialized and highly qualified capabilities but must encompass a full-spectrum force, the DPG argues. Hence, a key slogan for the new orientation's force structure is "breadth rather than depth" ("*Breite vor Tiefe*"), meaning a preference for "a little bit of everything" over further military specialization. This strategy incurs deficits in sustainability and effectiveness in operations but is said to give Germany a key political role in cooperating with European partners of small and medium size. By offering broad basic capabilities, Germany allows other partners to develop highly specialized forces that can then be pooled and shared in common operations — presumably, at times, under German leadership and with financial benefits for all.

In assessing this strategy and its translation into military reform, several problems stand out. For instance, with the end of conscription, it is yet unclear whether the envisioned troop strength will be sustainable, and at what cost. To maintain a force of 185,000 troops, about 12,500 new career and longer-term service members need to be recruited each year.[12] Given the rule of thumb that the Bundeswehr needs four applicants to fill one job satisfactorily, this is more of a challenge than it might seem at first glance. Early data on recruitment under the new system have been inconclusive.

Even if the ranks can be filled, the restructuring of the Bundeswehr into a rapidly deployable fighting force still stops half way. Of 185,000 troops, the government is only aiming to deploy a maximum of 10,000. That is a low level of ambition, even if one takes into account that to deploy 10,000, an additional 20,000 will be either in preparation and training to deploy or resting from a previous deployment. The political decision to limit each tour to just 4 months (instead of the more common 6 to 8 months) adds further pressure on personnel planning. Finally, it should be noted that having a capability to deploy 10,000 personnel is the defense ministry's stated goal; it remains unclear whether it will be achieved.

The idea of a force geared toward deployable operations abroad is not fully realized in terms of military hardware either. The Bundeswehr still lacks essential capabilities in areas such as tactical and strategic airlift. The proposed further reductions in helicopters and planned procurement of transport aircraft (A400M) do not mesh with the strategic analysis set out by the ministry, but they are a consequence of rising prices for new equipment and limited budgets.[13]

Especially in terms of capabilities, the current reform is designed very tightly, not allowing for much wiggle room for when a specific system runs into development problems or fails to materialize altogether. As the procurement process is notoriously unpredictable, this can thwart strategic planning — with serious consequences for German freedom of action. The most recent example of this is the cancellation of the unmanned aerial vehicle Euro Hawk because of licensing problems.[14] The ministry's decision, which came rather late in the procurement process, prompted a parliamentary investigation into whether money was wasted on a system that was known to be unfit. In the midst of a federal election campaign, that investigation received much attention, overshadowing the more central question of why Germany needs (armed and unarmed) unmanned aerial vehicles and how to fill this capability gap.

Beyond these issues of manpower, hardware, and procurement, there are also political problems. The rather ambitious role envisioned by the defense ministry for German armed forces in international security is lacking support from the public, parliament, and even parts of Chancellor Merkel's coalition government. Most Bundeswehr missions abroad are not supported by a majority of the German people. German support for the largest and most well-known mission, ISAF, has been dwindling for years, from 64 percent in 2005 to 44 percent in 2010 to 37 percent in 2011.[15] More consequential than this assessment of current or past missions is the deep reluctance to engage in similar operations again.

This tension is perhaps best encapsulated in foreign minister Guido Westerwelle's self-proclaimed doctrine, the *Culture of Military Restraint*, which is at odds with de Maizière's plea to take on "more military responsibility."[16] It is no accident that the DPG paper issued by de Maizière is only a ministerial one; its bold assignment of tasks to the Bundeswehr would most likely not be approved by Westerwelle's foreign ministry and would therefore not make it into a government-approved white book or similar statement by the German government as a whole. This lack of strategic consensus, of course, also affects the reform of the armed forces. In fact, it goes a long way in explaining the root causes of the problems outlined earlier.

These political divisions and the general desire not to repeat the Afghanistan experience point to a larger issue: Germany's political leadership is instinctively reluctant to use hard power. The use of military means is suspected to be rarely effective in producing desired political outcomes and always incurs political costs at home. As a nation deeply ashamed of the horrors of Nazi militarism and having been reeducated as free-riding consumers of security, Germany still struggles with the appropriate approach to military means as an instrument of foreign policy. Moreover, the average German does not feel threatened by turmoil abroad and sees little or no connection between safety at home and the need to maintain a stable liberal international order. It is little or no surprise, then, that so much of German foreign policy is predicated instead on trade, soft diplomacy, and on occasion, unilateral disarmament initiatives.

This combination makes Germany an unpredictable partner in international security affairs. There is always a chance that Germany's aversion to hard power will trump its strategic interests. Most prominently, that was the case in Libya in 2011 when Germany abstained in the United Nations Security Council—the first time it did not vote with France or the United States in that body—and subsequently refused to let its airborne surveillance capability (AWACS) contribute to NATO's Unified Protector mission.[17]

AWACS is a typical multinational capability, the very embodiment of the pooled and shared arrangements of "smart defense" that Germany keeps advocating in both the EU and NATO councils. Given such an example, it is not surprising that pooling-and-sharing arrangements are making little progress these days. This is not just a problem for Germany's and the EU's credibility as effective actors in international security, but also for the "new orientation" that is designed with a view to deeper European defense integration. The whole concept of "breadth rather than depth," for instance, will prove hollow without sufficient cooperation with others, especially in Europe.

CONCLUSION

Assessing Germany's hard power is a treacherous undertaking. There are two main reasons for this: first, in the midst of far-reaching Bundeswehr reform, all hard facts—from the eventual size of the force to actual capabilities—are uncertain and in flux. Minister de Maizière aims to complete his new orientation in 2017; until then, many of the numbers discussed here are goals or data whose programs are works in progress. While certain trends are discernible, their extrapolation is by no means reliable. After all, the Merkel government has undertaken several surprising reversals on defense issues already—for example, the sudden suspension of conscription or the unexplained retraction of the announced €8.3 billion in defense budget cuts.

Second, the development of German hard power over the last 10 to 20 years has been characterized by deep ambiguity, in terms of both posture and policy. This is a reflection of the two competing stories about the grand narrative of Germany's security policy. In describing this ambiguity, it is important to note that the topic of German hard power does not lend itself to a straight story of unmitigated decline. The study of German hard power is not the opening line of a bitter joke. It rests on a modest but solid base of steady budgets in recent years and acquisition programs that, while modest in scale, are technologically advanced. This cautiously positive assessment of German hard power gains particular traction in comparison to the developments of other European nations, large or small. In the conventional military balance among Europe's big three, for instance, Germany is catching up—although admittedly, this is due in no small part to the severe defense budget cuts in both France and the United Kingdom (UK).[18] While Paris and London command crucial capabilities that Germany does not—nuclear weapons, amphibious forces, aircraft carrier(s)—these high-value assets eat up much of their shrinking budgets, giving Germany an edge in other areas such as tanks (vis-à-vis the UK) and aircraft (vis-à-vis France).

THE GERMAN MILITARY IN AFGHANISTAN

German military involvement in the ISAF in Afghanistan epitomizes the ambiguities of German security policy discussed here. It can serve as an example for both narratives presented in the introduction: that of a strong and increasingly confident nation shouldering its share of the burden of upholding international stability and that of an indecisive nation pursuing a minimalist approach to its role in international security affairs because of its instinctive rejection of hard power means.

After September 11, 2001, Chancellor Gerhard Schröder declared Germany's "unlimited solidarity" with the United States, and it was in Germany's former capital where, in accordance with the international Bonn Agreement, the foundation for ISAF was laid. Schröder put his own chancellorship on the line when he combined his decision to send German armed forces to Afghanistan with a parliamentary vote of confidence. He won narrowly. Also, it was Schröder's defense minister, Peter Struck (a Social Democrat as well), who coined the enduring rationale for this mission of the Bundeswehr, reflecting a new reality in the age of globalized threats such as international terrorism: "Germany's security is also to be defended at the Hindu Kush."[19]

It is telling that such strong political backing was required for a relatively modest contribution: the initial number of German soldiers to be deployed to Afghanistan was a mere 1,200. Today, almost 12 years later, the size of the mandate encompasses 4,400 soldiers. These numbers indicate that Germany underestimated the difficulty of the challenge at hand and chose a strategy of minimalist incrementalism in dealing with it. This is also evident from the fact that German decisionmakers always emphasized the nonviolent nature of the Bundeswehr's job in the stable northern provinces of Afghanistan: networked security (Vernetzte Sicherheit), a German version of NATO's "comprehensive approach," was the key phrase, meaning that the armed forces did everything from painting schools to drilling wells, but would refrain from engaging the enemy. In fact, one of the German caveats in the NATO plan of operations for Afghanistan dictated that German soldiers were to shoot only in self-defense in face of an attack or imminent threat—after having yelled warnings in several languages. The German parliament's somewhat fanciful insistence on a clear separation between Operation ENDURING FREEDOM (understood as the bloody counterterrorism mission in which Germany could not participate) and ISAF (understood as the civilian reconstruction mission in which German soldiers participated as a kind of armed technical relief agency) underscored this general discomfort with hard power in action.

True to its reactive nature, German policy toward Afghanistan did not change until the deteriorating security situation in northern Afghanistan revealed a glaring gap between rhetoric and reality. In April 2010, Defense Minister Karl-Theodor zu Guttenberg was the first high-ranking German official who called the Bundeswehr's mission a war. Fearing the legal and political implications, he added "colloquially speaking." (Officially, "non-international armed conflict within the parameters of international law" remained the German phrase of choice.[20]) Around the same time, some of the caveats were dropped, and the extreme restrictions of the rules of engagement were abandoned. As it turned out, despite limited equipment—in tactical airlift and reconnaissance, for example—the Bundeswehr performed admirably against the insurgents.

Between January 2002 and July 2013, 54 German soldiers lost their lives in Afghanistan. Although Germany has never before experienced such high casualties, public reaction was muted. Arguably, this is a sign of what former German president Horst Köhler called the public's "benevolent indifference" toward its armed forces rather than an expression of general agreement with Germany's fight alongside its allies and the Afghan government. After all, when in September 2009 an American fighter jet responded to a German colonel's call, striking two fuel tankers captured by insurgents and killing more than 90 civilians in the process, Germany—8 years into the war—had its first intense public debate about military operations in ISAF. Former defense minister Franz Josef Jung; his deputy Peter Wichert; and the highest-ranking German soldier, Inspector General Wolfgang Schneiderhan, lost their jobs over the incident.[21]

The debate also highlighted increasing frustration with the perceived lack of progress in Afghanistan. Given the length and cost of the mission, Germany experienced the same kind of fatigue other allies did; strategic concern quickly turned to finding an honorable exit strategy. The changing face of ISAF was the main catalyst for this. The mission had been sold to the German public as a stabilization effort in which German forces would assist in Afghanistan's peaceful development toward democracy and prosperity; it was not advertised as a prolonged war against insurgents of dubious background and motivation.

Accordingly, NATO's decision to redeploy by 2014 was met with an audible sigh of relief in Berlin and in most other member states. As of yet, it is uncertain how many Western troops will remain in Afghanistan after that date—not as a fighting force, but as advisers in training the Afghan security forces. It is a testament to Germany's complicated relationship with its hard power that it was the first NATO state to specify an "after 2014" contingent of about 800 soldiers. While most observers applaud this bold commitment to alliance solidarity and Afghan stability, German officials keep their fingers crossed that this training mission does not evolve into a war-fighting operation once again.

Moreover, the direction of de Maizière's reform is sensible: focus on deployability, create leaner and more flexible forces, push for better cooperation among EU and NATO partners, and emphasize the need to be able to actually fight. So when the Atlantic Council states that "German military weakness is NATO's most significant problem," one could easily think of weaker and faster declining powers within the alliance — and more significant problems, too, such as diverging threat perceptions among most members.[22]

Still, there is something to the charge brought forth by the council and others. The numbers — stable as they may be — are not impressive for a state of Germany's size, location, wealth, and political power. They are, of course, even less impressive in comparison to the increases in the defense budgets of rising powers such as China, Brazil, and India. If Germany will one day be able to send 10,000 soldiers into combat abroad equipped with some of the remaining Leopard 2 tanks, or to deploy abroad a dozen brand-new Eurofighter Typhoons (a nonstealthy aircraft of disputed competitiveness), will it make much of a difference?

One of NATO's lessons from the 2011 war in Libya is that without U.S. support, the European allies, led by Britain and France, could not mount a sustainable campaign for lack of ammunition and planes, among other things.[23] Germany did not participate in that operation against one of the world's weakest militaries, but in terms of more effective air-defense suppression and close-air support, it would hardly have improved Europe's performance in any case.

113

That Germany did not even try to make a differ-ence in this UN-mandated NATO mission makes mat-ters worse. It is indicative of a disconnect between the country's strategic interests and its political will to use force. For every step forward toward a normalization of German security policy (Kosovo and Afghanistan), there is a step backward (utterly restrictive rules of engagement and Libya). When German armed forces are sent into international missions, it is usually, first and foremost, explained to the public as a necessary act of solidarity with Germany's allies. Although this is a good argument, it should never be a substitute for a lucid formulation of German interests and a clear-eyed analysis of the threat to be countered.

One emerging threat is the increasing weakness of the European states when it comes to their hard power capabilities and thus their ability to secure their own periphery, let alone their global strategic interests. Germany, as the undisputed economic powerhouse and central political actor in Europe, would be well advised to lead the charge in reversing this dangerous trend. This, however, would require much stronger leadership on German security policy than the coun-try has enjoyed over most of the last 2 decades.

ENDNOTES - CHAPTER 5

1. This chapter was originally published as an essay on October 8, 2013.

2. Up until the court's decision in July 1994, the Basic Law for the Federal Republic of Germany was understood as restricting the use of German armed forces for collective geographical secu-rity of NATO member states and the Western European Union (WEU). The court held that the government was free to deploy German troops in NATO or WEU operations "to implement reso-lutions of the Security Council of the United Nations" but that, to

do so, the government would need the approval of the German parliament (Bundestag) first. See Karl-Heinz Börner, "The Future of German Operations Outside NATO," *Parameters*, Spring 1996, pp. 62–72, available from *strategicstudiesinstitute.army.mil/pubs/parameters/Articles/96spring/borner.htm*.

3. See Gary J. Schmitt, "The West's 'Hard Power' Deficit," *The Los Angeles Times*, September 11, 2012.

4. Some in the ministry even make the argument that, compared to the 2010 budget, such nonsubstantial, nominal increases mask the fact that in real money, the savings until 2015 practically amount to the storied $8.3 billion.

5. See Federal Ministry of Defence, "The 2012 Defence Budget," August 14, 2012, available from *www.bmvg.de/portal/a/bmvg/!ut/p/c4/NYxBCo NADAB_tJFCL976iqqXJdYQg0kqa3b7ffVQhjkNDExw4 diEMeTrqDDAJF_qXZGicTlyOoSLXUqATJIlydjxXrpUZ-dN-0zLVQyblFJlfl_wfueK8NIDrvZ6wSyiBio/*.

6. See Federal Ministry of Defence, "Ressortbericht zum Stand der Neuausrichtung der Bundeswehr" (Interagency Report on the State of the Reorientation of the Bundeswehr), Bundesministerium der Verteidigung, May 8, 2013, pp. 13–16, available from *www.bmvg.de/portal/a/bmvg/!ut/p/c4/NYvBCsIwEET_aDcBRerNE-hSvvdh4S9sQVpqkrJt68eNNDs7AO8xj8Im1ye0UnFBObsUR7U-zn6QNT3AO8cuG6QqREb_FMJeKjfRYPc05eGsUnocrATjLDll-nWZgpzNUALWqVNr7T6R3-70-1qzUEfzb0fcIvx8gOBJaR2/*; and Bundeswehr, "Die Starke der Streitkrafte" ("Strength of Forces"), August 9, 2013, available from *www.bundeswehr.de/portal/a/bwde/streitkraefte/grundlagen/staerke/*.

7. The Joint Support Service (Streitkräftebasis) was created in 2000 as a distinct organizational unit, unifying the armed forces' logistical services at home and abroad. The Joint Support Service also includes the two universities of the armed forces (in Hamburg and Munich) and the the secret service agency (Militärische Abschirmdienst) of the Bundeswehr.

8. Brian McGrath, "NATO at Sea: Trends in Allied Naval Power," Chap. 4, this volume.

9. See Federal Ministry of Defence, "Information on the Personnel Structure Model (PSM) 185," March 7, 2012, available from *www.bmvg.de/portal/a/bmvg/!ut/p/c4/NYyxDoMwDET_yCFd-WnUrYulaqS2wBbCCVXAix4GlH98w9E56yzud6U0pu428Uwrs-FtOabqTrsMOwbh4SjTPKjKQphoWUPjBknjDtOAswZpeTlI1m9uZ9 nE0IY2DUg4qsVOjFaRCIQXQ5TBYpBmgyXWWburLVP_Z7eT-2bvrXnU3OvHyau6-0HVC1uPA!!/*;* and Federal Ministry of Defence, "Reorientation of the Budeswehr: The Implementation Plans," December 6, 2012, available from *www.bmvg.de/portal /a/bmvg/!ut/p/c4/NYzBCsIwEET_KJsiQvFmqQcvHgTReEvb-JV1sN2WzaS9-vOnBGXiXNwy8oZT9SsErRfYTvMD1 dOo2081rMIn6EWVE0rTEiZQpss8YNpwFMOYfU5SNpo5w-HM_G9D0kVF3KrJSYRCvUcwSRafdZJFiDA3gbNU2trL_ VN_64S6H27G27bW5wzLP5x8aaE_T/*.

10. This analysis is very much in line with similar statements by NATO in 2010 and by the EU in 2008. See North Atlantic Treaty Organization, *Active Engagement, Modern Defence: Strategic Concept for the Defence and Security of the Members of the North Atlantic Treaty Organization*, Lisbon, Portugal: NATO, November 20, 2010, available from *www.nato.int/nato_static/assets/pdf/pdf_publications/20120214_strategic-concept-2010-eng.pdf*; and European Union, *Report on the Implementation of the European Security Strategy: Providing Security in a Changing World*, Brussels, Belgium: EU, December 11, 2008, available from *www.consilium.europa.eu/ueDocs/cms_Data/docs/pressdata/EN/reports/104630.pdf*.

11. German Ministry of Defence, *Defence Policy Guidelines: Safeguarding National Interests — Assuming International Responsibility — Shaping Security Together*, Berlin, Germany: May 18, 2011, p. 10, available from *www.nato.diplo.de/contentblob/3150944/ Daten/1318881/VM_deMaiziere_180511_eng_DLD.pdf*. The German version does not say "a benchmark" but rather says "the benchmark."

12. "Ressortbericht zum Stand der Neuausrichtung der Bundeswehr," p. 24. Germany's armed forces consist of service men or women who intend to spend their whole professional lives in the military, others who sign contracts to serve from 2 to 20 years, and enlistees who agree to serve much shorter service periods of anywhere between 7 and 23 months. The figure of 12,500 annually required new recruits noted in the text does not include those enlisting for this shorter period.

13. For a critique of the "missing link" in official German strategy documents between the analysis of the international security situation and the recommendations for German military planning, see Nick Witney and Olivier de France, "Europe's Strategic Cacophony," *Policy Brief*, London, UK: European Council on Foreign Relations, April 25, 2013, p. 5.

14. Minister de Maizière cancelled the development of the Euro Hawk drone in May 2013, citing massive problems in obtaining flight permits for European airspace because of an insufficient automatic anticollision system and a lack of access to construction documents. According to the German air force, the necessary updates would amount to an additional cost of €500 to €600 million—a proposition unacceptable to the political leadership. The leading producing companies, Northrop Grumman and European Aeronautic Defence and Space Company N.V., dispute the ministry's assessment. For a chronicle of the Euro Hawk "scandal," see Helena Baers, "The Euro Hawk Scandal: A Chronicle," *Deutsche Welle*, April 6, 2013, available from *www.dw.de/the-euro-hawk-scandal-a-chronicle/a-16856113*.

15. See Thomas Bulmahn, Rüdiger Fiebig, and Carolin Hilpert, Sicherheits-und Verteidigungspolitisches Meinungsklima in der Bundesrepublik Deutschland (Security and Defense Policy and Public Opinion in the Federal Republic of Germany), Strausburg, Germany: Sozialwissenschaftliches Institut der Bundeswehr, May 2011, p. 6; and "Halten Sie die den Afghanistan—Einsatz der Bundeswehr fur Richtig?" ("Is Keeping the Army in Afghanistan Right?"), *Statistica*, January 2011, available from *de.statista.com/statistik/daten/studie/37270/umfrage/beteiligung-der-bundeswehr-am-einsatz-in-afghanistan/*. Interestingly, there are no reliable polls on this question for 2012–13. In the case of Afghanistan, however, the public's growing skepticism did not fully translate into political decisionmaking. With the exception of the radical leftist party, all six parties represented in parliament did, at least at one point, vote in favor of the ISAF mandate.

16. Guido Westerwelle, "Das amt macht mir freude" ("The Office Makes Me Happy"), *Stern*, July 11, 2013, available from *www.fdp.de/Das-Amt-macht-mir-Freude/4943c18747i1p452/index.html*; and Thomas de Maizière, "Warum Schweigt der Kerl? Hat der was zu Verbergen?" ("Why Is He Silent? Does He Have Some-

thing To Hide?"), *Die Welt*, May 27, 2013, available from *www. welt.de/politik/ausland/article116574251/Warum-schweigt-der-Kerl-Hat-der-was-zu-verbergen.html.*

17. After much embarrassment, Germany finally decided to send 300 additional soldiers to Afghanistan, thus freeing allied AWACS units to move from there to Libya.

18. In 2013, Germany's defense budget is at €33 billion; the UK and France have €37 billion and €32 billion, respectively. By 2015, Germany plans to reduce spending to €30 billion, the UK to €35 billion, and France to €30 billion. See Christopher Chantrill, "Public Spending Details for 2010," available from *www.ukpub-licspending.co.uk/year_spending_2010UKbn_12bc1n_30*; and Martial Foucault, "The Defense Budget in France: Between Denial and Decline," *Ifri Focus Stratégique,* December 2012.

19. R. Nicholas Burns *et al., Anchoring the Alliance*, Washington, DC: Atlantic Council, May 2012, p. vi.

20. For a biting critique, see Steven Erlanger, "Libya's Dark Lesson for NATO," *The New York Times*, September 3, 2011.

21. Steffen Heberstreit, "Peter Struck ist tot" ("Peter Struck Is Dead"), *Frankfurter Rundschau*, December 19, 2012, available from *www.fr-online.de/politik/ex-verteidigungsminister–peter-struck-ist-tot,1472596,21150806.html.*

22. "Tabu-Bruch: Guttenberg spricht von Krieg in Afghanistan" ("Breaking Taboo: Guttenberg Speaks on War in Afghanistan"), *Der Spiegel*, April 4, 2010, available from *www.spiegel.de/ politik/ausland/tabu-bruch-guttenberg-spricht-von-krieg-in-afghanistan-a-687235.html.*

23. For a succinct analysis of Germany's changing Afghanistan policy — and the political fallout of the Kunduz incident in particular — see Timo Noetzel, "The German Politics of War: Kunduz and the War in Afghanistan," *International Affairs*, Vol. 87, No. 2, March 2011, pp. 397–417.

CHAPTER 6

SOUTH KOREA: RESPONDING TO THE NORTH KOREAN THREAT[1]

Bruce E. Bechtol

KEY POINTS

- South Korea faces a clear, present, and evolving threat from North Korea, with Kim Jong-un showing no indication of moving away from his father's violent and corrupt policies.
- South Korea's response to the North Korean threat has been uneven, with increased capabilities in some areas but less than what is needed in others.
- A key issue facing the Republic of Korea (ROK)-U.S. alliance is command and control of allied forces during wartime on the Korean Peninsula. A combined operating force must continue to exist to ensure full readiness and capability.

When analyzing the readiness, capabilities, and future initiatives of ROK's military, one must take into account the unique geopolitical position in which the ROK government finds itself. There is no ambiguous set of threats for South Korea. Rather, the largest and most dangerous threat to the stability and security of the Korean Peninsula is obvious: the Democratic People's Republic of North Korea (DPRK).

It is for this threat that policymakers in Seoul must ensure their military is ready. Providing an adequate defense against this threat is the cornerstone of the ROK–U.S. alliance and the most important foreign policy issue between these two allies. As survival of the nation-state is the number one priority for any national leader, all other issues for Seoul will be ancillary as long as there is a DPRK.

Recognizing that the threatening behavior of its belligerent neighbor to the north is the key military issue for the ROK, it is important to analyze that threat to determine what the priorities of the South Korean military will be and how the threat will influence planning for the ROK–U.S. alliance. Since 2010, North Korea has conducted two violent military provocations: one with a submarine that sank a ROK naval ship and one that involved an artillery barrage against a South Korean island that killed both military and civilian personnel.[2] North Korea also conducted yet another nuclear test in February 2013.[3]

In addition, the DPRK has shown with a test launch conducted in mid-December 2012 that it is now capable (or close to it) of building a missile that can hit Alaska, Hawaii, or perhaps even the west coast of the United States.[4] Pyongyang also has the capability of targeting all of South Korea and most of Japan with its ballistic missiles.[5]

North Korea has also continued to advance the capabilities and numbers of its armored forces, long-range artillery forces, and special operations forces.[6] Finally, Kim Jong-un has shown no indication that he has any intentions except to carry on the violent and corrupt policies of his father, Kim Jong-il. This means, of course, that South Korea and the ROK–U.S. alliance must continue to prepare for the multifaceted North Korean threat for the foreseeable future.

INITIATIVES AGAINST THE NORTH KOREAN THREAT

Despite calls by the Roh Moo-hyun administration (2003–08) for a "balancer policy" — a policy that moved South Korea away from its traditional security ties with the United States to a more neutral or balancing role between the United States, Japan, and the old communist bloc of China, Russia, and North Korea — the fact remains that the primary issue for which Seoul must build its military capabilities and plan its contingencies is North Korea.[7] This process has been exacerbated by the fact that the threat the DPRK presents has evolved and become even more complicated in recent years.

Following the two violent provocations in 2010 already described, it became obvious that the South Korean government and military needed to take steps to counter future provocations from North Korea. As noted North Korean specialist Robert M. Collins has stated:

> Since the end of the Korean Conflict in 1953, the ROK–U.S. alliance has done a very good job of deterring against a war initiated by North Korea. The alliance has not done a good job of deterring North Korean provocations.[8]

Thus, the planning, policies, and procedures South Koreans initiated (and coordinated with their key ally in Washington) are very timely and needed now more than ever.

During April 2013, it was reported that the United States and South Korea had finalized a plan to respond more forcefully and appropriately to North Korean provocations.[9] This new "counterprovocation" plan will ensure that there is a speedy "response in kind" that still prevents escalation to all-out war. The existence of the plan was also made public in part, it seems, because Seoul and Washington wanted to both warn the North Koreans and reassure the South Korean populace.

In an earlier and equally important move, the South Korean military established a separate Northwest Islands Command. The establishment of the new command and the appointment of a commander with the autonomy to respond with necessary force in a timely manner under more liberal rules of engagement empower the South Korean military to respond more effectively to violent provocations the North initiates in the Northern Limit Line (NLL) area.[10]

Formally established in June 2011, the command was first headed by Lieutenant General Yoo Nak-jun, the commandant of the ROK Marine Corps, with a Marine major general as deputy commander and a staff that includes colonels from each of the ROK military services. Built around a division-sized joint unit, with the key contingents being the ROK Marine Sixth Brigade and the Yeonpyeong Defense battalion, the new command now has the ability to respond to North Korean attacks more effectively and rapidly. As such, ROK forces are now better positioned to deter and defend against North Korean provocations.[11]

The attacks in 2010 and the rhetoric from North Korea since have had the opposite effect of what Pyongyang likely wanted. If anything, DPRK behavior has strengthened South Korea's resolve to strike back against North Korean aggression.[12] The South Korean Navy is now on a heightened state of readiness in the NLL area—the demarcation line in the West (Yellow) Sea between the DPRK and ROK—and has been equipped with the best maritime equipment that the government can provide.[13]

As part of its support for these new initiatives, the United States also stepped up exercises and training with ROK forces in the West Sea, close to the NLL area.[14] Although much of the effort for counterprovocation deterrence has focused on the NLL, this is not the only area where readiness is being upgraded. For example, in June 2013, additional self-propelled air-defense missiles were assigned to front-line units near the demilitarized zone (DMZ).[15]

South Korea also faces a threat from the DPRK's advances in cyber and electronic warfare. In recent years, North Korea has engaged in a series of cyber and electronic warfare attacks against the South Korean military, government, businesses, and nonprofit entities.[16] In response, the Defense Ministry established a Cyber Policy Department in early-2013, and the National Intelligence Service announced that its third department would give greater attention to "monitoring of cyberspace and telecommunications."[17] The North Koreans reportedly have 3,000 to 4,000 personnel engaged in cyber warfare. To enhance the ROK's capability to counter this rather large and well-trained force, the Defense Ministry announced that it will be working with the United States to deter and defend against this emerging threat.[18]

Meanwhile, because North Korea used global positioning system (GPS) jamming on hundreds of commercial flights and maritime navigational units in South Korea during 2012 and 2013, Seoul increased its surveillance of North Korean electronic jammers. The Ministry of Science and Future Planning announced plans to set up a system that can track down the "attack point and impact of jamming attempts."[19]

The DPRK's missile program has grown in both numbers and capabilities. It poses a serious problem to both South Korea and Japan. In response to that threat, Tokyo acquired the land-based PATRIOT Advanced Capability-3 (PAC-3) from the United States, deployed the Standard Missile (SM-3) on its Aegis-equipped Japanese destroyers, joined the U.S.-led ballistic missile defense (BMD) system, and established the Bilateral Joint Operating Command Center at Yokota Air Base with the United States to provide a common operating picture of any missile threat.[20]

In contrast, South Korea has not, as of yet, done any of these things — though Seoul has begun to develop a less expensive and less capable BMD system of its own. Despite the considerable threat the DPRK's arsenal of missiles aimed at South Korea poses, as recently as May 2013, the South Korean Defense Ministry reiterated the government's intention not to participate in a joint U.S.-ROK missile defense effort, let alone the trilateral (Japan, U.S., and ROK) ballistic missile defense architecture suggested by the chairman of the U.S. Joint Chiefs of Staff, General Martin Dempsey, during a visit to South Korea in late April 2013.[21]

While keeping its distance from the kind of cooperation on missile defenses undertaken by Japan and the United States, South Korea is moving forward with its own missile defense upgrades; in a recent budget, the defense ministry indicated it intends to spend nearly 14 percent of its entire budget on improving its missile defense capabilities.[22] In 2012, for example, South Korea purchased two Green Pine land-based missile defense radars and, under new budget plans, recently announced it would acquire PAC-3s.[23] In addition, South Korea announced in June 2013 that it would equip its Aegis destroyers with the Standard Missile 6 (SM-6) for low-altitude defense against cruise missiles, unmanned aerial vehicles, and aircraft. The SM-6 is an upgrade to the SM-2s that were deployed on South Korean Aegis destroyers.

More ambitiously, Seoul plans to establish a Missile Destruction System by 2020. According to reports, the system will be designed to detect imminent North Korean missile launches and enable South Korea to strike missile sites before an attack can be carried out. According to South Korean sources, the system will involve "spy satellites, surveillance drones for monitoring and attack systems, including missiles, fighter jets and warships."[24]

Indeed, it appears that a key reason the United States and South Korea negotiated new, more lenient guidelines to the Missile Technology Control Regime (MCTR) in 2012 was to give the ROK the option of deploying longer-range missiles and more sophisticated drones to cover all of North Korea. Under the previous MCTR 2001 agreement, South Korean missiles were limited in range to no more than 186 miles. With the new accord, South Korean missiles will have a maximum range of 500 miles, which is sufficient to give them the capability of reaching any area of North Korea from launch points well south of Seoul and the DMZ.[25] Although the new agreement regarding missile range adds to Seoul's ability to target key nodes in the North, actually doing so would be both an expensive undertaking and a capability the United States already provides. In addition, it will do nothing to enhance badly needed improvements in ROK ballistic missile defense capabilities.

The fact remains that the missile defense systems currently deployed by the South Koreans are inferior to those currently deployed by the United States and

Japan. If the ROK had simply purchased the systems American experts recommended, such as the PAC-3 and SM-3, South Korea would be better prepared for a ballistic missile attack from North Korea. In addition, by joining a U.S.-led BMD system, the South Koreans would have access to the U.S. Navy's X-Band radar and the U.S. Army's land-based radar associated with the Terminal High Altitude Area Defense. The U.S.-led system links together the capabilities of detection and destruction systems around the globe and matches them up with mobile BMD platforms such as Aegis-equipped ships.[26] By going its own way when it comes to missile defense, the South Korean government is limiting its ability to defend itself and its citizens.

Cost Sharing and Repositioning U.S. Bases.

The cost for stationing U.S. forces in South Korea has been, and remains, an important issue in both South Korea and the United States. The perception of some in the United States, particularly members of Congress, has been that Seoul needs to do more to cover its "fair share" given the level of security the United States provides its ally from North Korean aggression. Americans see a South Korea that is now a thriving democracy and an economic powerhouse and expect the South Koreans to pay more of the cost for stationing U.S. troops there.[27] Conversely, many on the left in South Korea believe that their government is paying more than its fair share, arguing that American estimates that South Korea has been paying 40–45 percent of the basing costs are on the low side, and that South Korea is already paying more than the 50 percent of the costs for which Washington is calling.[28]

The accord governing South Korean payments is known as the "Special Measures Agreement" (SMA) and covers nonpersonnel stationing costs (NPSC), such as labor costs for South Korean employees working with U.S. forces, the purchase of logistics and supplies, and the construction of military facilities. The first SMA took effect in 1991, and South Korea's contribution levels have increased steadily as the costs associated with NPSC have grown predictably.[29] The last SMA was signed in December 2009, with Seoul and Washington agreeing that South Korea would pay 760 billion won (roughly $570 million at the time) for NPSC costs and Seoul also agreeing to cost hikes not to exceed 4 percent a year.[30] With the SMA set to expire in December 2013, Washington and Seoul had set the end of October as a deadline for reaching a new agreement. Talks in October did not result in an agreement but the Americans keep pushing for an SMA in which the South Koreans would pay 50 percent of the cost. However, a new SMA was negotiated in January 2014.[31]

Another important initiative is the Land Partnership Program, based largely on a 2006 agreement between Washington and Seoul to consolidate significantly the U.S. military footprint in South Korea (see Figure 6-1). The deadline initially set for completing the consolidation was 2012, but, given the scale of the endeavor, it is no surprise that the deadline has not been exactly met, and a large portion of forces north of Seoul are yet to be repositioned.

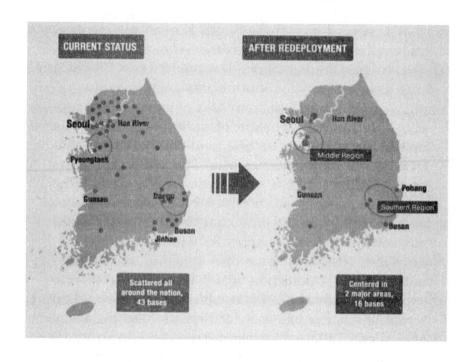

Source: Ministry of National Defense, ROK, "Defense White Paper," 2006.

Figure 6-1. Projected Relocation of U.S. Bases in South Korea.

Nevertheless, according to General James D. Thurman, commander of U.S. forces in Korea, "transitioning from 107 bases to less than 50" will ultimately result in "enhanced force protection, survivability, and lower cost maintenance in Korea."[32] The effect of this plan is already saving money for both the United States and South Korea.

BUDGETS AND ACQUISITIONS: PAYING FOR THE FUTURE

The Roh government in 2005 unleashed the most substantial reform agenda in recent years for the South Korean military, "Defense Reform 2020." This was the Roh government's vision for a ROK military that would be smaller, more modern, and capable of global missions—not just one focused on dealing with the North Korean threat. By 2020, the total military manpower would be cut by some 25 percent, with the ROK army seeing its numbers drop from 548,000 to 371,000—a loss of four corps and 23 divisions. These cuts were combined with reductions in the time conscripts would have to serve in the nation's army and navy by 6 months and in the air force by 8 months, with a deadline of 2014 for putting these new service requirements in place. In theory, these reductions in manpower would be made up with acquisition of new, advanced military hardware and systems.[33]

The plan, however, suffered from a number of problems. First, it required more resources than were budgeted. Second, many experts assessed the original schedule for systems acquisition and troop cuts to be inadequate to account for North Korea's own growing asymmetric capabilities in nuclear and ballistic missile weapons—a problem no doubt exacerbated by President Roh's overly sanguine view of North Korea's own strategic intentions. Third, the plan did not anticipate the command-and-control requirements that would flow from South Korea's decision to transition by 2015 to a more self-reliant force.[34]

Shortly after Lee Myung-bak was elected president in 2008, his government moved to modify both the substance and the timelines of Defense Reform Plan

(DRP) 2020. Taking the threat from North Korea more seriously, beginning in 2009, the ROK military reinforced plans to defend against the North Korean nuclear threat and to initiate troop cuts only after weapons systems have been brought online that would make up for the decrease in manpower.

Specifically, the revised plan, made public in 2009, included delaying the DRP 2020 reform endpoint to 2025, slowing defense budget increases as a result of slowdown in the Korean economy, and raising the planned 2020 troop level to 517,000 from the original goal of 500,000. The Lee government also modified the plan's reduction in service time for conscripts, with draftees in the army and the marines serving 21 months, navy conscripts 23 months, and air force draftees 24 months. Even so, the country's navy and air force are still likely to face manpower shortages in the coming years.[35]

The defense budget under President Roh began at 2.28 percent of gross domestic product (GDP) his first year in office. This percentage gradually went up and continued to go up after Lee Myung-bak assumed the presidency. Under Lee, it peaked at 2.72 percent of GDP in 2009 and was 2.60 percent in his last year in office.[36]

Before assuming office in February 2013, South Korean President Park Geun-hye stated that she intended to increase spending in light of Pyongyang's third nuclear test and its provocative behavior. In fact, her announced plan is to increase the defense budget at a higher rate than the overall state budget.[37]

In accord with those plans, the Defense Ministry announced in April 2013 that it intended to spend an extra $200 million during 2013—raising the 2013 budget from $30.5 billion to $30.7 billion. More recently, the ministry submitted a request to South Korea's legislature for a 2014-18 defense budget of $192.6 billion—an average annual expenditure of $38.52 billion. About half of 2013's increase was earmarked for strengthening defense capabilities along the ROK western maritime border with North Korea, and a bit less than half will be spent on upgrading existing conventional weaponry, such as South Korea's self-propelled 155 millimeter (mm) howitzers (K9 Thunder) and procuring additional unmanned reconnaissance aircraft.[38]

But challenges remain—as shown by the sinking of the ROKS *Cheonan* in March 2010 by a DPRK submarine. Increasing the ROK Navy's antisubmarine warfare capabilities should be a priority. Moreover, some key mainline battle systems need replacing, but replacements have been slow to come. One example is the K-2 Black Panther, an indigenously produced main battle tank intended to replace the American-made M-48 Patton tanks that the ROK Army still has in its inventory. (M-48s date from the 1950s and were the principal tank the U.S. Army used during the Vietnam War.) Mass production of the tank was originally set to begin in 2011, but the project was set back by numerous delays, including a failed engine durability test just in 2013.[39]

Also worrisome is the fact that South Korea's plan to buy 60 new fighter jets has been delayed. Only recently has the competition been reopened after all three of the entries—Boeing's F-15, Lockheed Martin's F-35, and European Aerospace Defense and Space Company's Eurofighter Typhoon—failed to fall below the price level set by the ROK's acquisition agency.[40] The country needs to replace its very old fleet of F-4 Phantoms and F-5 Tigers, and the F-35 would be the most advanced aircraft of the three — but also the most expensive. Whether South Korea's defense budget can accommodate such a purchase, whether offset proposals to reduce overall costs for the proposed acquisition can be arranged, or whether the government will simply be forced to buy fewer planes remain open questions.

WARTIME OPERATIONAL CONTROL: A KEY DEFENSE ISSUE

Since 1994, the Combined Forces Command (CFC) has had a planning staff of hundreds of ROK and U.S. personnel. The staff is commanded by a U.S. four-star general. During peacetime, ROK forces report to their relevant commands, which then answer to South Korea's Joint Chiefs of Staff. During wartime, designated ROK forces fall under the operational control (OPCON) of the commander of CFC, who in turn reports to the national command authorities in both Washington and Seoul. However, this long-standing agreement has been subject to intense negotiation and a number of proposed changes.

In 2007, Secretary of Defense Robert Gates and Defense Minister Kim Jang-soo reached an agreement that CFC would be disestablished, and the two militaries stationed in Korea would continue to function as allies but with two separate wartime operational commands. The new command architecture was to become operational in April 2012.[41]

The issue of American and South Korean forces fighting a conflict with North Korea under two separate military commands became an immediate source of contention in this new agreement. Senior politicians on the right and many retired military officers were highly critical of the change because they believed it

132

was both premature and dangerous to the security of South Korea.[42]

Under the current CFC structure, the military chain of command is transparent and seamless while falling under two separate national command authorities (NCA) in Washington and Seoul (see Figure 6-2). Although planning is conducted using a combined staff and exercises are held every year that utilize that planning, the ROK military does not "come under" the U.S. military even when CFC is activated because the American CFC commander answers to both NCAs.

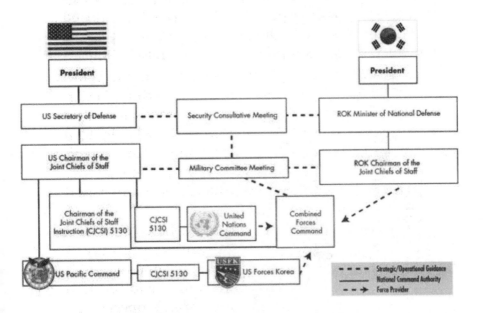

Source: Lieutenant General Stephen G. Wood, USAF, and Major Christopher A. Johnson, MD, USAF, "The Transformation of Air Forces on the Korean Peninsula," *Air and Space Power Journal*, Vol. 22, No. 3, Fall 2008, p. 6, available from *www.airpower.maxwell. af.mil/airchronicles/apj/apj08/fal08/wood.html*.

Figure 6-2. Current Wartime Command Relationships, ROK-U.S. Forces.

As originally conceived in 2008 and agreed to by Gates and Kim, the new command arrangement would no longer have ROK forces being put under the command of the CFC and its U.S. four-star commander. The CFC would no longer exist and, in its place, there would be two separate war-fighting commands—one American and one South Korean (see Figure 6-3). Unity of command, so important in war, would vanish, and U.S. and South Korean forces would be fighting in the challenging and restricted terrain of the Korean Peninsula, while answering to two separate NCAs. Much of the combined operations and planning today was slated to become cooperative through newly created boards, bureaus, coordination centers, and cells—a bureaucratic and complicated endeavor, to be sure.

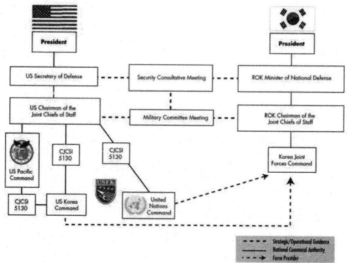

Source: Lieutenant General Stephen G. Wood, USAF, and Major Christopher A. Johnson, MD, USAF, "The Transformation of Air Forces on the Korean Peninsula," *Air and Space Power Journal*, Vol. 22, No. 3, Fall 2008, p. 7, available from *www.airpower.maxwell. af.mil/airchronicles/apj/apj08/fal08/wood.html.*

Figure 6-3. Projected Wartime Command Relationships Originally Slated for Post-2012.

In June 2010, Presidents Lee and Barack Obama agreed that the command changes would be delayed until December 2015.[43] This would give the ROK military more time to prepare for the types of planning and operations that separate warfighting commands would warrant; equally important, it would give the American and South Korean militaries time to modify and ameliorate some of the problems tied to the originally proposed command architecture.

Following Kim Jong-il's death and the accession of his son, Kim Jong-un, to the leadership of the DPRK in December 2011, events on the ground caused many in South Korea to again bring up the issue of the disestablishment of CFC.[44] North Korea conducted two long-range missile tests; staged another nuclear test; and, during the early spring of 2013, upped its level of threatening rhetoric.

As an editorial in a widely read South Korean newspaper put it:

> The South Korean government has proposed to the United States that the two allies reassess North Korean threats and the South Korean military's readiness posture ahead of the planned [change] . . . scheduled for December 2015. The proposal indicates that Seoul's security situation and its military's actual capabilities are more important than implementing the OPCON transfer on schedule. What is important, is that whether or not the OPCON transfer is implemented on schedule, the combined operational capabilities of the two allies' militaries for coping with threats from the North should not be weakened.[45]

But South Koreans were not the only ones to suggest the command reforms should be put on hold. In April 2013, former U.S. Forces Korea (and CFC) Commander General B. B. Bell argued that, in light of the DPRK's nuclear and missile capability, the change-over should be delayed to sometime past 2015—this from a general, who when CFC commander, had been a strong proponent of the change in command arrangements.[46]

Nevertheless, in April, the ROK defense ministry reiterated its intention to move forward with a new command structure and have it operational by the December 2015 deadline.[47] By early April 2013, reports had begun to circulate that, following the disestablishment of CFC in 2015, a new combined command would be stood up to take its place—essentially keeping the extremely important combined aspect of the ROK–U.S. alliance's fighting forces intact during wartime—though details were sketchy at the time (see Figure 6-4).[48]

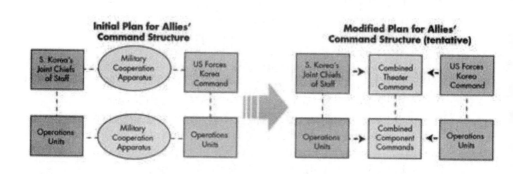

Source: Song Sang-ho, "Allies Agree on New Combined Command," *Korea Herald*, June 2, 2013, available from *www.koreaherald. com/view.php?ud=201306020000282*.

Figure 6-4. Projected ROK–U.S. Combined Command Structure Post-CFC, April 2013.

Although many details still needed to be worked out, in June 2013, it was reported that the new combined command would be headed by a ROK four-star, with an American general serving as deputy commander of the combined forces and an American air force general heading up the combined air component. By some accounts, ROK officers would command the other components.[49]

In July 2013, the South Korean government reportedly proposed to the United States that the originally agreed date for disestablishing CFC be once again delayed in light of the ongoing threat from North Korea. It is thus now unclear if "wartime OPCON" and the end of CFC will once again be pushed back to a date beyond 2015 or if the new combined command structure will, in fact, be implemented on that date.

According to press reports, in October 2013, U.S. Defense Secretary Chuck Hagel and South Korean Defense Minister Kim Kwan-jin agreed to delay the final decision until 2014.[50] What is most important for the future is maintaining a combined command that gives these two long-standing allies the optimum capability for combat readiness and deterrence of the North Korean threat.

THE U.S.-ROK NUCLEAR PACT

The United States and South Korea first signed a nuclear cooperation agreement in 1956, and it was last amended in 1974. With the accord set to expire in March 2014, Washington and Seoul have been in negotiations for over the past 2 years to extend and update the agreement. The main sticking point has been South Korea's desire to reprocess spent nuclear fuel of U.S. origin used in South Korean reactors — a practice effectively prohibited under the previous accord.

Unable to reprocess spent fuel, South Korea expects to run out of storage space for its spent fuel rods by 2016.[51] While Seoul has stated it wants to use "proliferation-resistant" technology for enriching uranium and reprocessing spent nuclear fuel, Washington has been hesitant to agree.

In light of North Korea's nuclear violation of the Nonproliferation Treaty and continuing nuclear program shenanigans, most states with an interest in the region are highly sensitive to any programs that might possibly increase the chances of weapons proliferation. Also, a likely issue for the United States is the past history of South Korea's own nuclear weapons program. Although Seoul had denied that it intends to engage in any effort that might lead it to acquiring nuclear weapons, recent polls show that a majority of the South Korean populace would support such an initiative.[52]

By March 2013, the United States and South Korea had failed to agree on how Seoul should (or should not) enrich uranium and process spent nuclear fuel rods. In talks held during June 2013, Ambassador Park Ro-byug from South Korea and Thomas Countryman from the United States continued to discuss the issues surrounding what Seoul would do with its "nuclear waste." As a temporary solution, the two countries have agreed to extend the existing accord by 2 more years, to March 2016. The 2-year extension of the present agreement must be approved by the U.S. Congress.[53]

Both countries hope to have reached a satisfactory compromise by then.[54] As long as the North Korean threat exists—and the perceptions about nuclear weapons that come with it—prospects for a South Korean reprocessing program will continue to be an issue. (It is important to note that while this book was in production the U.S. and South Korea signed a new nuclear energy cooperation pact on April 22, 2015.)

CONCLUSION

Since becoming an independent nation following the end of World War II, South Korea has never been more powerful on the world stage — militarily or economically. But the continuing unpredictable threat from North Korea means that South Korea must make significant investments in its national security.

South Korea needs to make important decisions regarding BMD; the future of its air force; numerous conventional systems that are vital to any conflict it would have with the DPRK; and, perhaps most important, the ROK–U.S. alliance and the command-and-control issues associated with the projected disestablishment of CFC in December 2015. These decisions are important, often quite expensive fiscally, and often very controversial politically. But this is nothing new.

South Korea is in a unique position. It is a thriving, transparent democracy, with perhaps the most ominous and imminent threat on its borders of any democracy. Decisions regarding the ROK military in coming years will be important to not only South Korea but also all nation-states that have an interest in the region.

ENDNOTES - CHAPTER 6

1. This essay was originally published on November 5, 2013.

2. For more analysis on the two violent provocations North Korea conducted against the South during 2010, see Alexander Zhebin, "The Korean Peninsula: Approaching the Danger Line," *Far Eastern Affairs*, No. 1, 2011, available from *www.eastviewpress*.

com/Files/FEA_FROM%20THE%20CURRENT%20ISSUE_No.%20
1_2011_small.pdf.

3. For details on North Korea's most recent nuclear test, see Kelsey Davenport, "North Korean Conducts Nuclear Test," *Arms Control Today*, March 2013, available from *www.armscontrol.org/ act/2013_03/North-Korea-Conducts-Nuclear-Test*.

4. "North Korea's Missile Programme," BBC News, April 12, 2013, available from *www.bbc.co.uk/news/world-asia-17399847*.

5. Duyeon Kim, "Fact Sheet: North Korea's Nuclear and Ballistic Missile Programs," Washington, DC: Center for Arms Control and Non-Proliferation, July 2013, available from *armscontrolcenter. org/fact-sheet-north-koreas-nuclear-and-ballistic-missile-programs/*.

6. For details on the recent updates, upgrades, and increasing numbers of North Korean conventional systems and forces, see Office of the Secretary of Defense, "Military and Security Developments Involving the Democratic People's Republic of Korea 2012," Annual Report to Congress, Washington, DC: Department of Defense, 2013, available from *www.defense.gov/pubs/Report_to_ Congress_on_Military_and_Security_Developments_Involving_the_ DPRK.pdf*.

7. For more analysis on the "Balancer Policy," see Yōichi Funabashi, *The Peninsula Question: A Chronicle of the Second North Korean Nuclear Crisis*, Washington, DC: Brookings Institution Press, 2007, pp. 252–257.

8. Robert M. Collins, Remarks at the "Confronting Security Challenges on the Korean Peninsula" conference, Marine Corps University, Quantico, VA, September 1, 2010. See also Robert Collins, "North Korea's Strategy of Compellence, Provocations, and the Northern Limit Line," Bruce Bechtol, ed., *Confronting Security Challenges on the Korean Peninsula*, Quantico, VA: Marine Corps University Press, 2011, p. 13, available from *community.marines. mil/news/publications/Documents/Confronting%20Security%20 Challenges.%20On%20The%20Korean%20Peninsula.pdf*.

9. David Sanger and Thom Shanker, "U.S. Designs a Korea Response Proportional to the Provocation," *The New York Times*,

April 7, 2013, available from *www.nytimes.com/2013/04/08/world/ asia/us-and-south-korea-devise-plan-to-counter-north.html*.

10. For more on the origins, history, and strategic importance of the Northern Limit Line and its role in the ongoing tense relationship between the two Koreas when it comes to the de facto sea border, see Terence Roehrig, "The Origins of the Northern Limit Line Dispute," Washington, DC: Woodrow Wilson International Center for Scholars, May 2012, available from *www.wilsoncenter. org/sites/default/files/NKIDP_eDossier_6_Origins_of_the_Northern_ Limit_Line.pdf*.

11. For information regarding the establishment and implementation of the Northwest Islands Defense Command, see "S. Korea Sets Up Defense Command for Yellow Sea Islands Near N. Korea," *Yonhap*, June 14, 2011, available from *english.yonhap- news.co.kr/national/2011/06/14/85/0301000000AEN20110614004500 315F.HTML*; and "Marines Recall Yeonpyeong Shelling with New Perspective," *Korea Times*, November 21, 2011, available from *www.koreatimes.co.kr/www/news/nation/2011/11/117_99175.html*.

12. Chico Harlan, "Island Attack Boosted S. Korea's Will to Strike Back against North," *The Washington Post*, April 14, 2013, available from *www.washingtonpost.com/world/asia_ pacific/island-attack-boosted-s-koreas-will-to-strike-back-against- north/2013/04/14/5d6a8a8c-a4d9-11e2-9c03-6952ff305f35_story.html*.

13. Song Sang-ho, "Navy Committed to NLL Defense," *Korea Herald*, July 1, 2013, available from *www.koreaherald.com/view. php?ud=20130701001036*.

14. Kim Eun-jung, "S. Korea, U.S. Hold Submarine Drill in Yellow Sea," *Yonhap*, May 6, 2013, available from *english.yonhap- news.co.kr/national/2013/05/06/16/0301000000AEN20130506003200 315F.HTML*.

15. "S. Korea Deploys Anti-Aircraft Missiles against N. Korea," *Yonhap*, June 24, 2013, available from *www.globalpost.com/dis- patch/news/yonhap-news-agency/130624/s-korea-deploys-anti-aircraft- missiles-against-n-korea*.

16. For examples of North Korean cyber and electronic warfare attacks against the South, see Choe Sang-hun, "Cyber Attacks Disrupt Leading Korean Sites," *The New York Times*, June 25, 2013, available from *www.nytimes.com/2013/06/26/world/asia/cyberattacks-shut-down-leading-korean-sites.html*; and Shaun Waterman, "North Korean Jamming of GPS Shows System's Weakness," *The Washington Times*, August 23, 2012, available from *www.washingtontimes.com/news/2012/aug/23/north-korean-jamming-gps-shows-systems-weakness.*

17. To read more about initiatives within the ROK Defense Ministry to combat cyber warfare attacks and efforts to work with the United States in this arena, see Kim Eun-jung, "S. Korean Military to Prepare with U.S. for Cyber Warfare Scenarios," *Yonhap*, April 1, 2013, available from *english.yonhapnews.co.kr/national/20 13/04/01/20/0301000000AEN20130401004000315F.HTML*; "Seoul Needs to Counter N. Korea's Cyber Espionage Capabilities: Defense Chief," *Yonhap*, June 20, 2013, available from *www.globalpost.com/dispatch/news/yonhap-news-agency/130620/seoul-needs-counter-n-koreas-cyber-espionage-capabilities-de*; Kim Tae-gyu, "Spy Agency Ups Capabilities against Cyber Attacks," *Korea Times*, April 12, 2013, available from *www.koreatimes.co.kr/www/news/nation/2013/04/116_133851.html*; and "Defense Ministry to Establish Cyber Policy Department," *Arrirang News*, April 2, 2013, available from *www.arirang.co.kr/News/News_View.asp?nseq=145526.*

18. *Ibid.*

19. Lee Minji, "S. Korea to Set Up GPS Jamming Surveillance System," *Yonhap*, April 10, 2013, available from *www.globalpost.com/dispatch/news/asianet/130410/s-korea-set-gps-jamming-surveillance-system.*

20. "Japan Deploys PAC-3 and SM-3 Missile Interceptors," *Defense Updates*, April 10, 2013, available from *defenseupdates.blogspot.com/2013/04/japan-deploys-pac-3-and-sm-3-missile.html*; and Ian Rinehart, Steven Hildreth, and Susan Lawrence, *Ballistic Missile Defense in the Asia-Pacific Region: Cooperation and Opposition*, Washington, DC: Congressional Research Service, June 24, 2013, available from *www.fas.org/sgp/crs/nuke/R43116.pdf.*

21. Karen Parrish, "Leaving Asia, Dempsey Discusses Combined Defense, China Engagement," *American Forces Press*, April 28, 2013, available from *www.defense.gov/news/newsarticle. aspx?id=119894*.

22. Kim Eun-jung, "S. Korea to Deploy Indigenous Missile Defense System in July," *Yonhap*, April 10, 2013, available from *english.yonhapnews.co.kr/national/2013/04/10/17/0301000000AEN20 130410010900315F.HTML*; and Zachary Keck, "South Korea Goes All In on Missile Defense," *Diplomat*, July 26, 2013, available from *thediplomat.com/flashpoints-blog/2013/07/26/south-korea-goes-all-in-on-missile-defense/*.

23. "PAC-2 Missiles Flunk Intercept Test," *Chosunilbo*, August 10, 2013, available from *english.chosun.com/site/data/html_dir/2012/10/29/2012102901030.html*; and Keck.

24. Kim Eun-jung, "South Korea Aims to Establish Missile Destruction System by 2020," *Yonhap*, June 11, 2013, available from *www.globalpost.com/dispatch/news/yonhap-news-agency/130611/s-korea-aims-establish-missile-destruction-system-2020-0*.

25. Jung Ha-Won, "U.S. Lets S. Korea Raise Missile Range to Cover North," Agence France Presse, October 7, 2012, available from *sg.news.yahoo.com/korea-us-agree-raise-missile-range-limit-054723718.html*.

26. For details regarding the American BMD system and the enhancements it would bring to BMD on the Korean Peninsula, see Loren Thompson, "Can U.S. Defenses Cope with North Korea's Missiles?" *Forbes*, April 5, 2013, available from *www.forbes. com/sites/lorenthompson/2013/04/05/can-u-s-defenses-cope-with-north-koreas-missiles/*. For details regarding the capabilities and deployment of the SM-2 and SM-6 BMD systems, see Kim Eun-jung, "S. Korea to Deploy New Surface-to-Air Missiles for Aegis Destroyers," *Yonhap*, June 12, 2013, available from *english.yonhap-news.co.kr/national/2013/06/12/37/0301000000AEN20130612004900 315F.HTML*.

27. See Chung Min-uck, "ROK, U.S., Differ in Troop Cost Sharing," *Korea Times*, June 4, 2013, available from *www.korea-times.co.kr/www/news/nation/2013/06/120_136899.html*.

28. "Civic Groups Call for End of U.S. Troops Cost-Sharing," *Yonhap*, July 2, 2013, available from *english.yonhapnews.co.kr/news/ 2013/07/02/0200000000AEN20130702008000315.HTML*.

29. "S. Korea, U.S. to Begin Defense Cost-Sharing Meeting in Washington," KBS News, July 2, 2013, available from *world.kbs. co.kr/english/news/news_Po_detail.htm?No=96978*.

30. Lee Chi-dong, "S. Korea to Pay 760 Billion Won for U.S. Troops in 2009," *Yonhap*, December 23, 2008, available from *english.yonhapnews.co.kr/news/2008/12/23/0200000000AEN200 81223007900315.HTML*.

31. See "Korea-U.S. Defense Cost Sharing Talks," KBS News, July 7, 2013, available from *english.kbs.co.kr/news/hot_is-sues_view.html?No=113154*; and "Seoul, Washington Fail to Agree on Defense Cost Sharing," *Yonhap*, October 31, 2013, available from *english.yonhapnews.co.kr/search1/2603000000. html?cid=AEN20131031005152315*; and "United States and Republic of Korea Finalize New Special Measures Agreement," Embassy of the United States, Seoul, Korea, January 12, 2014, available from *seoul.usembassy.gov/p_pr_011214.html*.

32. See Statement of General John D. Thurman before the House Armed Services Committee, March 28, 2012, available from *www.usfk.mil/usfk/Uploads/110/Statement.pdf?AspxAutoDetect CookieSupport=1*.

33. For details about Joint Vision 2020, see Bruce Klingner, "South Korea: Taking the Right Steps to Defense Reform," *Heritage Foundation Backgrounder #2618 on Asia and the Pacific*, October 19, 2011, available from *www.heritage.org/research/reports/2011/10/ south-korea-taking-the-right-steps-toward-defense-reform*.

34. Jung Sung-ki, "Defense Reform Faces Overhaul," *Korea Times*, August 27, 2008, available from *www.koreatimes.co.kr/www/ news/nation/2008/08/116_30141.html*.

35. For details of requested ROK Defense Ministry changes to its projected budget plans, and the slowdown of troop cuts, see Jung Sung-ki, "Less Spending for Military Modernization," *Korea Times*, April 7, 2009, available from *www.koreatimes.co.kr/*

www/news/nation/2009/04/113_42785.html; Jung Sung-ki, "S. Korean Military to Slow Down Troop Cuts," *Korea Times*, November 24, 2008, available from *www.koreatimes.co.kr/www/news/nation/2008/11/205_34989.html*; and Jung Sung-ki, "Service Period Cuts Will Fan Manpower Shortage in Military," *Korea Times*, October 9, 2008, available from *www.koreatimes.co.kr/www/news/nation/2008/10/113_32457.html*.

36. "S. Korea's Defense Spending Rises Amid N. Korea's Nuclear Threat," *Yonhap*, April 7, 2013, available from *english.yonhapnews.co.kr/national/2013/04/07/91/0301000000AEN20130407000700315F.HTML*.

37. Park Byong-su, "Incoming Park Administration Plans to Increase Defense Spending," *Hankyoreh*, February 22, 2013, available from *www.hani.co.kr/arti/english_edition/e_national/575126.html*.

38. "S. Korea to Increase Defense Spending by $200 Million This Year," *Arrirang News*, April 16, 2013, available from *www.arirang.co.kr/News/News_View.asp?nseq=146064*.

39. Kim Eun-jung, "S. Korea's K2 Battle Tank Fails Another Engine Test," *Yonhap*, April 22, 2013, available from *english.yonhapnews.co.kr/national/2013/04/22/23/0301000000AEN20130422007100315F.HTML*.

40. See Kim Tae-gyu, "Will Korea Barter F-35 for T-50?" *Korea Times*, May 9, 2013, available from *www.koreatimes.co.kr/www/news/nation/2013/05/205_135445.html*; Kim Eun-jung, "S. Korean Fighter Jet Project Stuck Over Pricing," *Yonhap*, July 8, 2013, available from *english.yonhapnews.co.kr/fullstory/2013/07/08/98/4500000000AEN20130708004600315F.HTML*, and "South Korea Reopens Bidding for 8.3 Trillion Won Fighter Jet Competition," *Reuters*, July 25, 2013, available from *www.reuters.com/article/2013/07/25/korea-fighter-bidding-idUSL4N0FV1RK20130725*.

41. For details of the signed agreement between Secretary Gates and Minister Kim, see "Secretary Gates Holds Consultations with ROK Minister of National Defense," *Defense Link*, February 23, 2007, available from *www.defenselink.mil/news/Feb2007/d20070223sdrok.pdf*.

42. "Former Generals Criticize Seoul-Washington Deal on Wartime Control Transfer," *Yonhap*, February 26, 2007, available from *english.yonhapnews.co.kr/Engnews/20070226/610000000020070 226140348E0.html.*

43. Victor Cha, "U.S.-Korea Relations: The Sinking of the Cheonan," *Comparative Connections*, Vol. 12, No. 2, 2010, available from *csis.org/files/publication/1002qus_korea.pdf.*

44. Kim Tae-gyu, "Wartime Command Transition Now in Doubt," *Korea Times*, March 10, 2013, available from *www.korea-times.co.kr/www/news/nation/2013/03/116_131848.html.*

45. "National Security Is Key in Determining Delay in OP-CON Transfer," *Donga Ilbo*, July 18, 2013, available from *english. donga.com/srv/service.php3?bicode=080000&biid=2013071868238.*

46. See Ashley Rowland, "Former USFK Commander Speaks Out Against Giving S. Korea Operational Control," *Stars and Stripes*, April 29, 2013, available from *www.stripes.com/news/pacific/ former-usfk-commander-speaks-out-against-giving-s-korea-operational-control-1.218742.*

47. See Song Sang-ho, "Calls against Wartime Control Transfer Resurface Amid Tension," *Korea Herald*, April 22, 2013, available from *www.koreaherald.com/view.php?ud=20130422000911*; and Kim Eun-jung, "Seoul Vows to Keep Deadline for Wartime Control Transfer," *Yonhap*, April 22, 2013, available from *english.yon-hapnews.co.kr/national/2013/04/22/42/0301000000AEN20130422003 451315F.HTML.*

48. "Transfer of Wartime Control Unclear in Military's Plans," *Hankyoreh Ilbo*, April 2, 2013, available from *www.hani.co.kr/arti/ english_edition /e_national/580872.html.*

49. For details regarding the many aspects of the new projected structure as of June 2013, see Song Sang-ho, "Allies Agree on New Combined Command," *Korea Herald*, June 2, 2013, available from *www.koreaherald.com/view.php?ud=20130602000282.*

50. "Troop Control Handover is a Vital and Delicate Issue," *Chosun Ilbo*, October 4, 2013, available from *english.chosun.com/site/ data/html_dir/2013/10/04/2013100401664.html.*

51. Hwang Sung-hee, "Korea, U.S. Agree to Extend Current Nuclear Pact by 2 Years," *Arrirang News*, April 25, 2013, available from *www.arirang.co.kr/News/News_View.asp?nseq=146394*.

52. *Ibid.*

53. Mark Manyin *et al.*, *U.S.–South Korea Relations*, Washington, DC: Congressional Research Service, April 26, 2013, available from *www.fas.org/sgp/crs/row/R41481.pdf*.

54. "S. Korea, U.S. Fail to Bridge Gaps on Reprocessing, Uranium Enrichment," *Yonhap*, June 4, 2013, available from *english.yonhapnews.co.kr/national/2013/06/04/1/0301000000AEN201306040 07700315F.HTML*.

CHAPTER 7

POLISH HARD POWER: INVESTING IN THE MILITARY AS EUROPE CUTS BACK[1]

Andrew A. Michta

Andrew Michta would like to thank his research assistants, Jacob Foreman and Matthew Washnock, for their contribution to this chapter.

KEY POINTS

- Unlike America's other major European allies, Poland's growing economy has allowed it to increase its defense spending.
- Warsaw's strategic focus has increasingly turned to improving Poland's territorial defenses and working with neighboring allies to bolster regional security.
- Poland has begun a major military modernization program whose success will depend on the continued health of the Polish economy and the transformation of the Polish defense industry into an efficient producer of advanced military equipment.

Poland's security strategy rests on the twin pillars of the North Atlantic Treaty Organization (NATO) and the European Union (EU). As the American military presence in Europe continues to shrink, however, Poland's support for the EU has increased, benefitting from EU structural-fund transfers, expanded trade, and integration under the Schengen Agreement. Consequently, while NATO and the United States remain essential to Poland's security, today Germany is Poland's key ally on the continent, with Polish public opinion showing for the first time in a 2012 survey a preference for Germany over the United States.[2]

Though positive attitudes toward the United States rebounded somewhat a year later, clearly the Polish public has become more distant in its view of America. The Barack Obama administration's 2009 decision to cancel the George W. Bush–era missile shield whose ground interceptors were to be based in Poland was a shock to bilateral ties. Announced on the 70th anniversary of the 1939 Soviet invasion of Poland, it became a public relations debacle for Washington. Compounding problems is the administration's more recent decision to scrap its plans for deploying high-speed Standard Missile 3 Block IIB interceptors in Poland and Romania (Phase Four of the European Phased Adaptive Approach) and Washington's continued reluctance to lift the visa requirement for Poles travelling to the United States. While there remains a large reservoir of public goodwill in Poland toward the United States, which has a large Polish-American ethnic community and a history of close military cooperation in recent years, these decisions have chipped away at traditional pro-U.S. sentiments in Poland.

Similarly, while Poland remains committed to NATO as the military pillar of its national security and, as such, a strong supporter of NATO's Article V tasks of collective defense, it has also become more vocal in support of the EU Common Security and Defense Policy. Again, while the United States remains Poland's principal ally and the country has been an active participant in American-led operations—with the largest being in Iraq and Afghanistan—there has been a marked decline in public support for current and future expeditionary missions, as exemplified in Warsaw's decision to not join other NATO allies in Operation UNIFIED PROTECTOR, the 2011 Libyan military campaign.

Poland's increased focus on Article V matters is tied largely to its growing concern about the resurgence of Russia's power and influence along Poland's eastern border. Since eastward NATO enlargement, especially to Ukraine, has all but vanished from U.S. and European security policy agendas, Poland finds itself in a border-state position within the alliance. Warsaw's perception of a changing regional power balance has brought about a new emphasis on the defense of national territory in Poland, making Warsaw refocus its attention closer to home as it plans to adapt the armed forces accordingly.

Over the past 5 years, Poland has focused more and more on its indigenous national defense capabilities, with the government funneling resources for military modernization. Because of its history of foreign invasions, the country has a keen appreciation of the vital importance of a strong military to the nation's sovereignty and security. An old Polish saying captures well the public mood on national defense: "If you can count, ultimately count on yourself."

BUCKING EUROPEAN TRENDS

Amidst the current protracted economic crisis in Europe and despite a 2013 slowdown in growth in Poland's own economy, Poland remains one of the EU's most dynamic countries. Today, it is its ninth-biggest economy, having increased by almost one-fifth since 2009.[3] Because the government is required under Polish law to spend 1.95 percent of its annual gross domestic product (GDP) on defense, a growing economy has allowed Warsaw to buck the general European trend of cutting national defense budgets (see Figure 7-1).

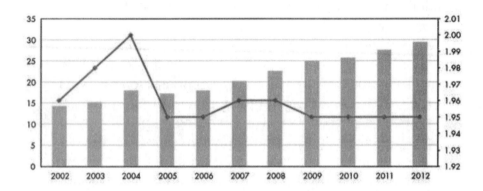

Source: Ministry of National Defence, Republic of Poland, "Basic Information on the MoND Budget, 2001-12," available from *archiwalny.mon.gov.pl/en/strona/126/lg_89.*

Figure 7-1. Total Defense Spending (Billions of Polish PLN) and Defense Spending as a Percentage of GDP.

With increased resources, Poland's ministry of defense has launched "The Modernization Plan for the Armed Forces in the Years 2013–2022" — the country's most ambitious program to date, which will include new ships, helicopters, tanks and armored personnel carriers, additional aircraft, and most importantly, new air and missile defenses.[4] The antiballistic missile (ABM) system is the most significant of Poland's military modernization efforts in terms of planned dedicated resources. The estimated cost of Poland's ABM program is set between $4 and $6 billion, making it the largest acquisition program in the country's history.

In mid-2013, however, with the economy slowing, Polish Prime Minister Donald Tusk was forced to revise the government's budget, resulting in a 10 percent cut to the defense budget.[5] Despite these reductions, Minister of Defense Tomasz Siemoniak has emphasized that the country's strategic projects will be protected, announcing in late September 2013 that military modernization will reach PLN 91.5 billion (approximately $30 billion) through 2022, covering 14 specific programs.

Consistent with Poland's desire to develop its military capabilities, the Polish government has renewed its focus on modernizing and expanding the country's indigenous defense industrial sector. In the fall of 2013, the government began the process of consolidating Poland's defense industry into a unified Polish Defense Group (Polska Grupa Zbrojeniowa [PGZ]) with the expectation that it would improve the sector's efficiency and competitiveness. The PGZ will combine the flagship Polish Defense Holding [Polski Holding Obronny, formerly Bumar] with Huta Stalowa Wola, among others. The effort has just begun, so it is too early to judge its ultimate impact on the industry. But the decision indicates the seriousness of the government's commitment to modernizing the defense sector and to making it more competitive in international markets.

The immediate question going forward will be whether the Polish military can still leverage available resources and complete the key elements of the modernization program despite the 10 percent budget decrease. Since it is government policy that modernization be done through the Polish defense industry whenever possible, there will be considerable focus on whether those firms can, in fact, deliver the product the military needs, and especially whether they can partner with foreign firms to leverage synergies with the domestic sector. In short, will Poland manage to continue committing enough resources to remain one of the few countries in Europe that is still serious about military power, and thereby become a NATO ally with growing capabilities and political clout?

MILITARY MODERNIZATION PLANS

Poland has doubled its defense spending over the past decade. Initially, the government budgeted PLN 31.4 billion on defense (approximately $10 billion) for 2013. Even with the planned 10 percent reductions in the 2013 defense budget, there has been a significant infusion of resources into the Polish armed forces. The current military modernization plan calls for spending PLN 91.5 billion through 2022 and stipulates that PLN 16 billion will be expended by 2016. The government has also restated that maintaining 1.95 percent of GDP on defense remains a priority.

As part of the modernization process, Poland began establishing two new high-level military commands starting January 1, 2014.[6] The goal is to create a joint operational command by replacing the separate service commands, converting them into departments, and turning the general staff into a strategic planning and advisory command.

The government also intends to maximize the use of the Polish defense industry with "Polonization" of the defense modernization effort tied to technology transfer from international partners as acquisition plans move forward. In addition, the government plans to spend PLN 40 billion on purchases not included in the 2014–22 operational plans. In total, Poland plans to spend approximately PLN 139 billion ($U.S.46.3 billion) on equipment modernization across the services, on added information technology capabilities, and on increasing the overall combat readiness of the Polish forces. In the process, Poland plans to build its modernization effort around 14 major programs.[7] Considering the scope of programs and resources allocated, a significant challenge for the defense ministry will be to improve the acquisition process to ensure platforms and equipment are fielded; in previous years, the ministry has even returned funds to the state budget.

For 2013, the Polish ministry of defense planned to increase capital expenditures to 26.2 percent of the budget—a 4.2 percent increase compared to the previous 3 years (see Figure 7-2).[8] The structure of the current Polish defense budget reflects the ministry's commitment to reverse the current approximate one-to-three ratio of modern-to-legacy military systems. Polish military equipment remains a mix of Soviet-era legacy systems (sometimes adapted with Western equipment) and innovative Polish designs developed in cooperation with Western firms.

For example, Polish land forces maintain 901 main battle tanks, of which 128 are the older-generation German Leopard 2A4s, 232 are PT-91 Twardys (a Polish modification of the Soviet T-72), and 541 are obsolete T-72s of three different types. Likewise, Poland maintains a fleet of 1,784 armored infantry fighting vehicles (AIFV), of which more than two-thirds are legacy Soviet BMP-1s, but nearly 500 are the highly capable KTO Rosomak, a Polish version of a Finnish AIFV that has been battlefield tested in Afghanistan.

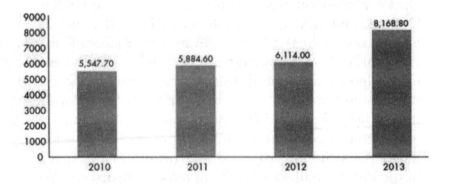

Source: Ministry of National Defence, Finance Department, *Podstawowe informacje o budzecie resort obrony narodowej na 2013* (Basic Information on the Ministry of National Defense Budget in 2013), Warsaw, Poland, March 2013, available from *mon.gov.pl/z/ mon.gov.pl/z/pliki/dokumenty/rozne/2013/09/informator_o_budzecie_ resortu_ON_na_2013_r..pdf.*

Figure 7-2. Increase in Procurement Expenditures (Missions PLN).

To help address this problem, however, in November 2013 Poland signed an agreement to purchase from Germany an additional 105 Leopard 2A5s, plus 14 Leopard 2A4s and 200 support vehicles.[9]

Addressing deficiencies in air mobility also remains a priority, as Polish military helicopters are currently a combination of Soviet-era systems and the aging PZL Sokół platform and its derivatives. To do so, the army will be seeking to acquire up to 70 new helicopters. The defense ministry also plans to issue funds for new modular armored vehicles, unmanned aerial vehicles (including armed drones), self-propelled howitzers, heavy mortars, antitank missiles, and new communication equipment.

The Polish navy has five tactical submarines (four German-built, 1960s-era Kobben class and one Soviet-legacy Kilo), two principal surface combatants (Oliver Hazard Perry–class frigates), a corvette (Polish-built ORP Kaszub class), and a number of mine warfare, mine countermeasure, patrol, amphibious, and support ships. The navy's aviation element includes two naval aviation bases, with equipment deployed in three locations. Two of those locations are home to air groups that include planes and helicopters for transport, antisubmarine, and search-and-rescue operations. The navy's modernization program includes new patrol boats, minesweepers, coastal-defense vessels, and possibly up to three submarines.[10]

Of the three major services, the Polish air force ranks as the most modern among post-communist states of Central Europe, averaging 160–200 flying hours per year (comparable to France's and exceeding Germany's). The air force operates three squadrons of F-16C/Ds, two squadrons of MiG-29A/UBs, and two squadrons of fighter/ground-attack Su-22M-4s. The Sukhoi aircraft have been slated for removal from service, and Poland will be looking to purchase additional Western planes or unmanned aerial vehicles. Two air force transport squadrons fly a combination of C-130E, C-295M, and Polish PZL M-28 Bryza aircraft. The air force also operates two squadrons of transport helicopters which, as noted previously, are aging platforms.

On balance, the most successful air force program so far has been the addition of F-16 jet fighters to its fleet of aircraft, accelerating the modernization process and increasing NATO interoperability. A visible sign of progress has been the opening of a U.S. training facility in the central Polish town of Łask for rotational exercises of U.S. and NATO aircraft.

Air and missile defenses (AMD), however, remain Poland's top defense priority. A law Poland passed this year appears to guarantee stable funding for the systems.[11] The program will combine a medium-range missile and air defense system and a variety of shorter-range systems with plans to expand the coverage for the country's entire territory. The government will allocate PLN 26.4 billion for AMD through 2022, with PLN 1.2 billion planned for 2014–16.[12]

Overall, Poland's shopping list is extensive; some would call it overly ambitious. While the air and missile defense budget seems protected, in light of the slowing economy and this year's reduction in planned defense expenditures, there is already talk of reducing the number of helicopters in the initial order and of cuts in other procurement programs. Indeed, there are also questions as to whether—even if all the acquisition programs were fully funded—Poland's defense ministry would be able to meet its acquisition plans. Some analysts have pointed out that based on the current track record of procurement, and especially the rate of contract fulfillment in 2012, Poland may again have a shortfall from the original spending plans.[13]

LEVERAGING DEFENSE FOR INDUSTRIAL CAPACITY

The Polish government sees military modernization as a path to modernizing the country's defense industry. The increase in procurement funds has attracted a lot of attention from U.S. and European defense industries—something the Polish government is determined to leverage for national defense industry modernization. Until 2013, Poland spent between 15 to 22 percent of its defense budget on equipment modernization. Poland's expeditionary missions in both Iraq and Afghanistan have highlighted the need for a better equipment kit for its forces, and the current program ultimately aims to shift about one-third of the defense budget to equipment modernization over the next decade.

Here, the AMD project is seen as central not just to the national defense strategy but also to preserving and expanding Poland's indigenous defense industrial capacity. Defense Minister Tomasz Siemoniak has repeatedly made clear that any AMD solution adopted by the government will need to involve extensive cooperation with Polish defense companies. It must include both long-term partnerships and significant technology transfers.

The army expects the initial components of the system to be tested in 2017 and a working system capable of defending national territory from an attack is to be in place by 2023—all procured with the direct participation of the Polish defense sector. For the Polish defense industry, this is a once-in-a-lifetime opportunity to partner with the best Western firms. Eventually, the government hopes to shift up to 80 percent of future work on particular defense projects to Polish suppliers.

One aspect of Polish military modernization seldom discussed is its intra-EU political dimension. As Poland undertakes its military modernization effort and defense ministry officials push for the maximum possible participation of Polish firms in plans to buy missiles, ships, helicopters, tanks, and small arms, it will run up against the growing pressure within the EU to reduce national preference in defense contracts.[14]

The planned purchases also seek to leverage domestic industry on smaller ticket items such as the

MSBS 5.56 program to develop a new modular assault rifle for Polish forces and the Tytan program comprising a system of technologies, similar to the U.S. Land Warrior, to be used by an individual soldier.[15] This effort to maximize domestic industry participation applies to both equipment upgrades and new system purchases; however, it may meet serious obstacles considering the imbalances of expertise and capacity in the Polish defense sector, as seen in the delays in modernizing Poland's Leopard 2 tanks.

The extent to which Polonization is likely to work will be best tested on high-end systems. There will be mounting pressure to give as much of the ABM work as possible to Polish companies.[16] Initial competition for the AMD contract is already underway with U.S., French, and Israeli systems expected to emerge as the principal contenders. But the key question for Polish officials is likely to be: Which of the foreign contractors can best coordinate with Polish defense firms to build a long-term and mutually beneficial partnership?

STRATEGIC PRIORITIES

Poland's level of defense spending and new acquisition programs reflects growing concern about the changing geostrategic environment in Central Europe following two landmark developments: The 2008 Bucharest NATO summit that, for all practical purposes, ended prospects of NATO membership for Ukraine and Georgia, and the 2008 Russo-Georgian war that brought back the specter of conventional state-on-state conflict along Europe's periphery. NATO's refusal to offer Ukraine a Membership Action Plan, combined with Russia's growing geostrategic assertiveness, has forced Poland to revisit traditional dilemmas associated with being a boundary state along the frontier of the West. More than anything else, Russia's invasion of Georgia drove home the critical importance of having workable NATO contingency plans and sufficient capabilities to perform key national defense tasks to make those plans credible.

The *Defense Strategy of the Republic of Poland*, adopted in 2009, captures both the enduring principles and the changing context of Poland's strategic thinking.[17] While NATO and the United States remain central to Poland's security, there has been a reorientation in Poland's strategy leading to an emphasis on regional and traditional territorial defense tasks over the past 5 years. Warsaw would like to keep relations with Washington close, and military and intelligence cooperation between the American and Polish militaries remains exemplary, with the Poles having accumulated a wealth of experience working closely with the United States in Iraq and Afghanistan.

Nevertheless, there is a sense within Poland of a growing "transatlantic deficit" in ties between the United States and its NATO allies in Central Europe, with the United States being seen as increasingly absent from the region. In particular, the Obama administration's decision to cancel both the George W. Bush administration's plans for antimissile deployments to Poland and its own plans to do the same—along with its 2012 decision to reduce the number of American forces based in Europe—has led Poland to give more attention to its own strategic and military options should the American security guarantee grow even weaker.

While the Polish government remains committed to NATO as the core pillar of its national security, Poland is also looking for greater regional security cooperation among the Nordic, Baltic, and Central European states to bolster its own security plans. Warsaw is also actively seeking to reenergize the Weimar Triangle (Poland, France, and Germany) and the Visegrád Group (Poland, the Czech Republic, Hungary, and Slovakia). Although Poland considers the possibility of a large-scale conflict with Russia unlikely, Poland has increasingly focused on the potential of local conflicts with states close to its border.[18] Here, the militarization of Russia's Kaliningrad enclave in the northeast has become a major issue.

Although Poland shares alliance-wide concerns about cyber and other nontraditional security issues, regional geostrategic considerations remain paramount to how the country approaches national security. Most importantly, while Poland continues to invest in regional security cooperation, it has made it clear that better regional ties should never come at the expense of allied solidarity or weaken the NATO-wide Article V security guarantee.

In 2013, Poland's National Security Bureau [Biura Bezpieczeństwa Narodowego], an advisory body to the country's president, published a comprehensive review on Poland's strategic position.[19] Without naming Russia as an outright foe, the white paper reflects Warsaw's growing preoccupation with resurgent Russian power as one of four key variables defining Poland's security (the other three being NATO, the United States, and the EU). Though not ruling out the possibility that Russia might choose a path of cooperation with the West, Poland's strategists have been skeptical about Russia's willingness to abandon its imperial aspirations, especially in light of reports that Russia has threatened to deploy 9K720 Iskander missiles in Kaliningrad and Moscow's actions in the post-Soviet "near-abroad."[20]

The relationship between the two countries has been further complicated by the aftermath of the Smolensk plane crash in 2010, which killed then-president Lech Kaczyński, his wife, and more than 90 of Poland's most senior military and political leaders. Continuing problems with Russia during and after the investigation of the crash, including Moscow's refusal to return the black boxes and wreckage of the Polish aircraft, have caused further friction between the two countries and remain an important domestic political issue in Poland.

Although few in Poland would argue that there is an imminent threat of aggression from Russia, Poles continue to see Russia as the principal threat to Poland's security and sovereignty. For this reason, some analysts have even suggested that if NATO solidarity continues to weaken, Poland will need to seek bilateral security agreements with the United States and Germany.[21]

Analysts have also been considering creating an improved conventional deterrent posture at the national level by mixing defensive and offensive systems, and adapting planning accordingly. To that end, Poland has closely followed the approach taken by the Finns, exploring the option of equipping its F-16s with stealth AGM-158 Joint Air-to-Surface Standoff Missiles. Another consideration has been the possibility of purchasing tactical ballistic missiles for its Multiple Launch Rocket System launchers and other systems that would give Poland medium- and possibly long-range strategic strike capability.[22]

Both the 2009 *Defense Strategy of the Republic of Poland* and the 2013 white paper reflect an evolving consensus on defense policy. The 2009 paper emphasizes the core importance of the dual pillars of NATO and EU membership for Poland's security. Recognizing the broadening array of nonstate and unconventional threats, the strategy paper emphasizes the core importance of balancing collective defense and international crisis response. The 2013 white paper recommends an approach that combines ongoing efforts to "internationalize" Poland's security within the existing alliance structure to ensure that an attack on Poland would generate a collective allied response. Finally, the paper seeks to place Polish strategic priorities in a larger context, with uncertainty surrounding the future of the EU and with declining American involvement in Europe—all pointing to the increasing need for Poland to become self-reliant in security matters, commensurate with the country's economic and military potential.

POLAND'S MILITARY ABROAD

Poland has a strong military tradition, a reputation it has lived up to in Iraq and Afghanistan. Poland's expeditionary missions in Iraq in support of Operation IRAQI FREEDOM and in Afghanistan as part of the International Security Assistance Force (ISAF) have been instrumental in shaping today's Polish armed forces.

Poland was an early participant in the 2003 Iraq military operation to oust Saddam Hussein, sending a small contingent at the start of the war and 2,500 troops for security and stability operations after the fall of Baghdad. Soon thereafter, on September 3, 2003, Poland assumed leadership of one of two multinational divisions and responsibility for a region covering five provinces. The core of the Polish-led divisions consisted of three brigades: Polish, Ukrainian, and Spanish, with military contingents and personnel from 24 other countries. Over time, the composition of the division changed with different countries offering contributions and others withdrawing their contingents. The mission evolved as well, changing from a post-conflict stability and reconstruction operation to one of combat and providing local security. Over time, the number of Polish troops deployed decreased from 2,400 to 900, with the last Polish troops withdrawing from Iraq in 2008.

On balance, Poland's participation in the Iraq mission gave the armed forces invaluable experience, laying the foundation for much of the country's current modernization plans. On the political side of the ledger, however, public support for the mission rapidly declined as Poles, contrary to expectations, saw few reconstruction projects in Iraq go to Polish firms and the security situation in Iraq worsened in the immediate aftermath of the invasion. In the end, Iraq inaugurated a new, more complex phase in U.S.-Polish relations.

As Poland pulled out of Iraq, it increased its contribution to the ISAF mission. At its peak, Poland deployed 2,600 soldiers to Afghanistan, at one point assuming responsibility for the entire Afghan province of Ghazni. The mission in Afghanistan was ultimately on an order of magnitude more challenging than the deployment in Iraq, both in terms of the threat environment and logistical difficulties. The Polish military is largely responsible for the mission's success, having adapted both personnel and equipment to the task. As the ISAF mission winds down, the key challenge for the Polish army is to repatriate and refurbish its equipment currently deployed in Afghanistan. Lacking indigenous capabilities for long-range lift, Poland will rely on the United States to facilitate the return of Polish equipment.

As with the Iraq mission, however, the Afghanistan operation has witnessed dwindling public support. This was especially true after the Obama administration decided to scrap deployment to Poland of the antiballistic missile system, and Poles began to question whether the sacrifices their military was making in Afghanistan and before that in Iraq were duly appreciated in Washington. As a result, Polish support for expeditionary operations has declined precipitously, as has overall public confidence in NATO's value to Poland's security. Polling data from a 2013 report by the German Marshall Fund of the United States suggests that when citizens of various NATO nations were asked whether NATO is still essential to their respective countries' security, Poles are 11 percentage points behind the EU average.[23]

In late-2013, Poland had approximately 1,940 soldiers deployed on various missions abroad, with the largest contingent deployed under ISAF in Afghanistan, followed by a contingent with the Kosovo Force, troops with the EU Force in Bosnia and Herzegovina, and a number of United Nations observers in Western Sahara, the Congo, Afghanistan, Kosovo, Liberia, South Sudan, and Côte d'Ivoire. Following the French campaign in Mali, Poland has also deployed trainers there. In addition, there are Polish military observers as part of the EU Monitoring Mission in Georgia. The total number of Polish military troops deployed outside of Poland was expected to decline further at the end of 2014 as the ISAF mission shifted from a combat to a support role.

CONCLUSION

Poland is, by any measure, the most successful case of post-communist political and economic transition to market democracy in Europe. As a relatively new member to NATO, it has made significant contributions to American and NATO military missions.

But Poland is entering an era of increasing uncertainty. America's commitment to European security appears to Poland to be waning, while Russia's resurgence as a military power in the context of Europe's de facto disarmament and the economic crisis within the EU raise even greater questions about Poland's future security environment.

To meet these challenges, Poland has clearly been an outlier among European NATO allies when it comes to national defense. Simply put, it is one of the few remaining European states serious about investing in its military despite the current economic crisis. As noted earlier, the primary focus of Poland's 10-year defense modernization plan is territorial defense rather than out-of-area capabilities, though Poland tries to balance the two with planned capabilities important to both, such as command, control, communications, computers, and intelligence as well as helicopter lift.

Two key questions loom over modernization plans. The first is the potential risk associated with the desire to use Polish defense companies to carry out the bulk of the modernization effort. There is no question that giving the lion's share of the work to Polish companies has great potential benefits for industrial modernization and employment, and employment is no doubt important to the government in Warsaw as Poland approaches its next parliamentary election in 2015. However, the record of the Polish industry has been spotty, with program delays and cost overruns.

The government seems aware of the risk. It has pushed to initiate the consolidation of the industry parallel with the modernization effort, as the Polish defense sector gears up for its largest contracts to date. However, the challenge will be to remain realistic about what can be achieved in the near term, recognizing that some of these companies face a steep learning curve when it comes to the kind of advanced manufacturing and systems engineering required to produce first-rate, up-to-date equipment. The key will be successful partnering with top international defense firms in a way that brings about transfers of manufacturing technology and has Polish companies focusing on those parts of the program where they are most competitive. Most importantly—and politically difficult—the government will need to be prepared for a course correction in its plans should Polonization of the modernization effort not deliver equipment and weapons platforms on time and in sufficient quantities. While domestic industrial priorities are important, they cannot overshadow the strategic requirements of the Polish Armed Forces.

The second question is whether the Polish economy will continue to grow at sufficient rates to sustain steady defense spending allocations to make the programs a reality. The 2013 cuts were not crippling for the Polish modernization effort, but if the government fails to stick by the 1.95 percent of GDP formula its ambitious program will need to be revised. The squeeze already seen in the defense budget should serve as a warning sign for the government that cutting defense—though politically seemingly less toxic than cuts in public spending—will eventually damage Poland's procurement plans and ultimately the nation's security. Hence, it is the 2014 state budget rather than the modifications to 2013 spending that will serve as a clear indicator of whether Poland remains serious about defense modernization.

With an economy that has performed better than its European neighbors, a desire to bolster and modernize its military capabilities, and a record of commitment to the transatlantic alliance, Poland continues to buck the trend when it comes to America's continental security partners. And with increasing influence in the EU, Poland continues to rise in the ranks as a midsize power and, as such, grow its potential to play an even greater role in Western security affairs in the future. But the budget decisions and program choices Poland makes in the next year and over the next decade will go a long way to determining just how great a role it will in fact play.

ENDNOTES - CHAPTER 7

1. This chapter was originally published as an essay on December 19, 2013.

2. German Marshall Fund of the United States, *Transatlantic Trends 2012*, Washington, DC, p. 11, available from *trends.gmfus. org/files/2012/09/TT-2012-Key-Findings-Report.pdf.*

3. In 2012, Poland's GDP was PLN 1.6 trillion, $530 billion. See Ministry of Treasury, "Macroeconomic Analysis of Polish Economy," December 13, 2013, available from *www.msp.gov.pl/en/ polish-economy /macroeconomic-analysis/4983,Macroeconomic-Analysis-of-Polish-economy.html.*

4. "Program rozwoju Sił Zbrojnych RP w latach 2013—2022" ("Program for the Development of the Polish Armed Forces, 2013-2022"), *Polska Zbrojna*, December 12, 2012, available from *www.polska-zbrojna.pl/home/articleshow/5744?t=Minister-okreslil-kierunki-rozwoju.*

5. In 2011, Poland's GDP grew 4.5 percent but is expected to expand by just 1.1 percent in 2013. See "Economic Activity Accelerates on Stronger Exports," *FocusEconomics*, August 30, 2013, available from *www.focus-economics.com/en/economy/news/Poland-GDP-Economic_activity_accelerates_on_stronger_exports-2013-08-30*. The planned 2013 defense budget amounted to PLN 31 billion. The revised budget scaled defense spending back by PLN 3.3 billion. Poland will be spending approximately the same amount as it did on defense in 2012.

6. Remigiusz Wilk, "Poland Creating New Military Commands," *IHS Jane's Defence Weekly*, September 17, 2013, available from *www.janes.com/article/27152/poland-creating-new-military-commands*.

7. Krzysztof Wilewski, "Pewne pieniądze na nowe uzbrojenie" ("Money for New Equipment Available"), *Polska Zbrojna*, September 17, 2013, available from *www.polska-zbrojna.pl/mobile/articleshow/9575?t=Pewne-pieniadze-na-nowe-uzbrojenie*.

8. Ministry of National Defence, Budget Department, *Podstawowe informacje o budzecie resortu obrony narodowej na 2013 (Basic Information About the Budget for the Ministry of National Defense in 2013)*, Warsaw, Poland, March 2013), available from *mon.gov.pl/z/pliki/dokumenty/rozne/2013/09/informator_o_budzecie_resortu_ON_na_2013_r..pdf*.

9. Ministry of National Defence, "Kolejne Leopardy dla wojsk lądowych" ("More Leopard for Ground Troops"), November 22, 2013, available from *mon.gov.pl/aktualnosci/artykul/najnowsze/2013-11-21-leopardy-dla-wojsk-ladowych/*.

10. "Renowacja Floty" ("Renovation of the Fleet"), *Polska Zbrojna*, July 24, 2013, available from *wiadomosci.wp.pl/kat,1020223,title,Renowacja-floty-do-2030-roku-Marynarka-Wojenna-RP-otrzyma-ponad-20-nowych-okretow,wid,15835580,wiadomosc.html?ticaid=111a17*.

11. "Poland Starts Work on Missile Defense System," *NewsEurope*, May 11, 2013, available from *newseurope.eu/2013/05/11/poland-starts-work-on-missile-defense-system/*.

12. "Rzad uchwalil program finansowania modernizacji wojska" ("The Government Has Adopted a Plan for Financing the Modernization of the Army"), *ONET.biznes*, September 17, 2013, available from *biznes.onet.pl/rzad-uchwalil-program-finansowania-modernizacji-wo,18512,5577553,news-detal.*

13. Zbigniew Lentowicz, "Armia musi mocno zacisnac pasa" ("The Army Has Tightened Its Belt"), *Rzeczpospolita*, July 19, 2013, available from *archiwum.rp.pl/artykul/1212341-Armia-musi-mocno-zacisnac-pasa.html.*

14. There is a growing realization in the EU that national preferences in defense contracting reduce efficiency. For a good summary of the debate, see Joachim Hofbauer *et al.*, *European Defense Trends 2012: Budgets, Regulatory Frameworks, and the Industrial Base*, Washington, DC: Center for Strategic and International Studies, 2012, pp. 43–46, available from *csis.org/files/publication/121212_Berteau_EuroDefenseTrends2012_Web.pdf.*

15. Marek Kozubal and Zbigniew Lentowicz, "Budujemy zolnieza przyszlosci" ("We Are Building the Future Soldier"), *Rzeczpospolita*, June 7, 2013, available from *archiwum.rp.pl/artykul/1191346-Budujemy-zolnierza-przyszlosci.html.*

16. Jan Cienski, "Defense: Polish Military Prepares for Modernization," *Financial Times*, May 22, 2013, available from *www.ft.com/intl/cms/s/0/d46e4d06-bbba-11e2-82df-00144feab7de.html#axzz2k5kIzEAY.*

17. All references to the document are drawn from the official translation issued by the Ministry of National Defence in 2009. See Ministry of National Defence, Defense Strategy of the Republic of Poland, *Sector Strategy of the National Security Strategy of the Republic of Poland*, Warsaw, Poland, 2009, available from *www.isn.ethz.ch/Digital-Library/Publications/Detail/?lng=en&id=156791.*

18. *Ibid.*, p. 5.

19. See National Security Bureau, "Biala Ksiega Bezpieczenstwa Narodowego Rzeczpospolitej Polskiej" ("White Paper on Polish National Security"), Warsaw, Poland, 2013, available from *www.spbn.gov.pl/sbn/biala-ksiega/4630,Biala-Ksiega.html.*

20. If true, the deployment of Iskander to the Kaliningrad district would change how Poland, the Baltic states, and Scandanavia look at their regional security. See Steve Gutterman, "Russia Has Stationed Iskander Missiles in Western Region: Reports," *Reuters*, December 16, 2013, available from *www.reuters.com/ article/2013/12/16/us-russia-missiles-idUSBRE9BF0W020131216*.

21. Andrzej Talaga, "Pytania, ktorych nie wolno zadac" ("Questions That Can't Be Asked"), *Rzeczpospolita*, May 24, 2013, available from *blog.rp.pl/talaga/2013/05/24/pytania-ktorych-nie-wolno-zadac/*.

22. Tomasz Szatkowski, "Polish Defense Modernization in the Era of U.S. Strategic Rebalancing," Washington, DC: Center for European Policy Analysis, March 1, 2013, available from *www.cepa.org/content/polish-defense-modernization-era-us-strategic-rebalancing*.

23. German Marshall Fund of the United States, *Transatlantic Trends: Topline Trends 2013*, Washington, DC, September 18, 2013, p. 24, available from *trends.gmfus.org/files/2013/09/TT-TOPLINE-DATA.pdf*.

CHAPTER 8

FRENCH HARD POWER:
LIVING ON THE STRATEGIC EDGE[1]

Dorothée Fouchaux

KEY POINTS

- With its 2013 defense white paper, France reaffirmed its intent to maintain its strategic autonomy by dint of its nuclear deterrent and by retaining a conventional power-projection capability.
- To carry out this program and do so while facing budget constraints, French forces will continue to decline both in total numbers and numbers deployable.
- Meeting the white paper's goals rests on potentially overly optimistic assumptions about program savings, export offsets, and future European defense cooperation.

Before the publication of France's latest defense white paper in April 2013, French newspapers were predicting a virtual "tsunami" in cuts to the country's defense budget and force structure.[2] Although the finance ministry hoped to use savings from a greatly reduced defense budget to help bring the country's public deficit down to less than 3 percent of its gross domestic product (GDP), Defense Minister Jean-Yves Le Drian, key members of the French Parliament, and the French defense industry lobbied French President François Hollande to stave off deep cuts to the military.[3] Then, in a televised speech a month before the white paper's publication, Hollande said defense would not face greater budget reductions than any other government ministry.

According to the appropriations statute that follows and implements the white paper's program, the French defense budget would flat line at €3.38 billion over the next 2 years and creep ever so slowly to €32.51 billion in 2019. A decline, to be sure, from the resource expectations set out in the 2008 white paper—but not as precipitous as some had predicted.[4]

Although not as confident sounding as the 2008 white paper with regard to France's ability to meet the security challenges of the current year, the 2013 white paper nevertheless maintains the country's core strategic ambitions by protecting the defense budget in three areas: "autonomy in decision-making, protection of the French territory, [and] nuclear deterrence."[5] The question, however, is whether even after fending off more serious cuts, there remain sufficient resources for France to retain its capacity to field an adequately sized, fully trained, and modernized force that can meet those strategic goals.

Indeed, there is already a gap of approximately €45 billion between the military's past plans and resulting budgets.[6] Should the French economy continue to lag, there will be pressure again to look to the defense budget for additional savings. In short, is the 2013 white paper a realistic assessment of the future of French defense capabilities, or does it signal the start of a subtle but noticeable decline in the country's strategic ambitions?

TRANSFORMATION IN AN ERA OF DECLINING RESOURCES

Since the end of the Cold War, France has, like other major Western powers, set about reforming its armed forces to meet the challenges of the new security environment. In 1994, a government white paper sought to create a plan for the military that no longer focused on dealing with a threat posed by the Soviet Union, but instead was directed at dealing with pockets of instability around the globe, the increased risk tied to the proliferation of weapons of mass destruction, and the appearance of asymmetric threats such as terrorism.

To maintain strategic relevance, the government reasoned that, while it needed to maintain its nuclear forces as a hedge against the threat of proliferation and to support French foreign policy independence, France required a new model for its armed forces. The Model 2015 (as it was then called) was to be:

> a professional, more compact army, better equipped and better adapted to actions outside the national territory. Its capacities were defined so as to allow, simultaneously, the development of permanent arrangement of prevention, a visible and significant presence in an international coalition, as well as more limited operations under national command, while providing the protection of the territory and its approaches.[7]

When Jacques Chirac came to power in 1995 as the French President, the government decided to end peacetime conscription and create an all-professional armed force. The goal was to form a military that would be readily deployable; could operate modern, complex weapon systems; and was capable of operating within an international coalition. This was a fundamental transformation of the French military both in terms of capabilities and size. According to the International Institute for Strategic Studies' annual *Military Balance*, France's active duty forces in 1997 totaled 358,800 (203,200 army, 63,300 navy, and 78,100 air force); in 2002, the numbers were 259,050 (137,000 army, 44,250 navy, and 64,000 air force).[8] In 2012, French active duty personnel had shrunk to 228,850 (122,500 army, 38,650 navy, and 49,850 air force).[9] By 2020, the expectation is that the military's active duty numbers will decline even further, dropping to approximately 190,000.[10]

At the same time that France was moving to an all-professional force, the government launched several major acquisition programs. These included the Tiger attack helicopter; the NH90 multirole helicopter; the armored infantry combat vehicles (VBCI); the nuclear-powered Barracuda-class attack submarines; and surface-to-air missile platform/terrain, a theater antimissile defense system. During this period, France also introduced into its fleet Europe's largest warship, the nuclear aircraft carrier *Charles de Gaulle*; a new generation of nuclear-powered ballistic missile submarines (SSBNs); and launched two Helios 1 optical surveillance satellites.

Though France's defense spending as a percentage of GDP started to decrease during that time, it dropped even further between 1997 and 2002, when France was governed by a coalition led by the Socialist Party. In 2002, the defense budget dropped to €28.85 billion (excluding pensions)—the lowest total since the end of the Cold War.[11] In addition to cuts in training and procurement, research and development (R&D) funding decreased by some 30 percent between 1997 and 2002. As in other Western countries, cuts in defense spending were used by the government as a means to reduce the public deficit.

The third white paper was released in summer 2008 following Nicolas Sarkozy's election as president the year before.[12] The paper, which purports to rest on a "strategic appraisal for the next 15 years," highlights the threats posed by cyber warfare, transnational actors, and nuclear proliferation. It puts special emphasis on increasing French intelligence capabilities to meet France's evolving security needs. It also announced the continued downsizing of defense personnel (civilian and military) by 54,900. To be carried out over a 6-year period, the downsizing was intended to free up monies to spend on new modernization programs for France's conventional and nuclear forces and the continuation of existing acquisition programs such as the army's Fantassin à Équipement et Liaisons Intégrés (Integrated Infantryman Equipment and Communications [FELIN]) infantry combat system and the navy's multirole frigate program (FREMM). The paper also set a goal for the French government of it being capable of deploying 30,000 soldiers abroad, with necessary air and naval support forces, for 1 year.

The economic crisis that followed the issuance of the 2008 white paper, however, made it fiscally challenging for the government to meet the paper's goals. As Figure 8-1 shows, the difference between planned and actual procurement expenditures had risen to more than €3 billion between 2009 and 2012. Several factors explain this decline, including the unexpected cost of operations in Libya in 2011, the expenditures related to creating the French military base in Abu Dhabi, and France's reintegration into the North Atlantic Treaty Organization's (NATO) military command structure. In addition, the government expected to reap more savings than occurred with the previous downsizing of the French military and civilian defense workforce.[13] Consequently, in 2012 the defense ministry decided to postpone €5.5 billion in procurement to help bring the budget back in line with existing resources.[14]

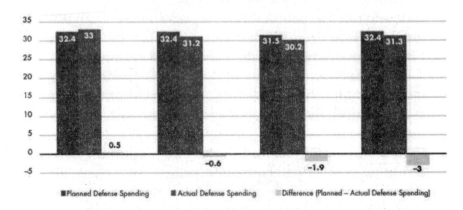

Source: French National Assembly, *rapport d'information No. 1388 (Information Report No. 1388)*, September 18, 2013, p. 20, available from *www.assemblee-nationale.fr/14/pdf/rap-info/i1388.pdf*.

Figure 8-1. Planned Spending and Actual Spending (Billions €).

A SHRINKING MARGIN OF DEFENSE

The next defense white paper was published in 2013. Though originally only intended to be an update of the 2008 white paper, the global financial crisis, Arab Spring, American pivot to Asia, and French intervention in Mali necessitated significant changes, resulting in a new document. The new paper also provided the recently elected President François Hollande an opportunity to put his own stamp on French defense policy.

France had not had a Socialist president in nearly 2 decades, and the last Socialist government was perceived as particularly difficult for the French military. Despite Hollande's statements to the effect that France needed to provide for its own security and maintain its nuclear deterrent, the defense community's memory of the previous Socialist government combined with the ongoing economic crisis led many to expect the worst. On its face, however, the 2013 white paper was not a major break from its 2008 predecessor. Nevertheless, because the document calls for further reduction in forces, argues for resizing the geographic region in which French military interventions would be legitimate, and indicates that military resources will be divvied up depending on the readiness and operational requirements of particular military units, the white paper's broader implications require more analysis.

In addition to eliminating 24,000 employees from the current staffing of the defense ministry, including troops and civilians (a figure that increases to nearly 34,000 when the 10,000 planned but still unexecuted cuts from 2008 are factored in), the white paper provides for a reorganization of the armed forces on the basis of what it calls "the principle of differentiation." Although exact details on the principle's implementation were not provided by the paper, it is described as "giving priority to the equipment and training" of some elements of the armed forces versus others. When combined with the effort to save additional monies by financing "costly or cutting-edge capabilities only when they are indispensable and benefit, in particular, forces set up to combat state-level actors,"[15] the two initiatives will undoubtedly have an impact on the state of the French military going forward.

The white paper appears to suggest that there will be a two-tiered system for the armed forces: one well-equipped and trained, the other slated for domestic security missions not requiring sophisticated or costly equipment.[16] The military personnel involved in domestic operations will have fewer opportunities to participate in operations abroad and will train with equipment that is less than state-of-the-art. On the whole, this makes the French Army less attractive as a profession, and could lead to major problems operationally should those troops be required to conduct operations abroad.

A second major concern generated by the white paper is its implications for military procurement. In March 2013, France's largest defense firms wrote a letter to Hollande expressing their concerns that when it comes to possible cuts in defense spending, "it is essential that industrial and socio-economic issues be taken into account as seriously as budget issues."[17]

The defense budget is perceived differently than other elements of public spending. French defense firms are seen as a pillar of French industry, providing high-skilled jobs and generating technological innovations that are of use in both the military and civilian domains. The defense industry also contributes positively to the country's balance of trade: one-third of its annual revenue, nearly €15 billion, comes from defense-related exports.[18]

To square the circle of saving money but maintaining France's defense industrial base, Hollande decided to continue procurement of most major weapons systems but simultaneously renegotiate the contracts for those systems by either buying fewer allotments or delaying deliveries and payments. Defense companies were compensated for the renegotiation of these contracts with firm orders in the amount of €45.2 billion (see Table 8-1).[19] While the French military remains one of the best-equipped militaries in the world in terms of the systems themselves, there are increasing worries as to whether they will be fielded in operationally relevant numbers.

Programs	2008 Planned Orders	Firm Orders
A400M aircraft	50	50
Rafale aircraft	286	180
Barracuda-class submarines	6	3
FREMM frigates	11	11
ASTER missiles	575	535
Naval cruise missiles	200	200
NH90 helicopters	160	61
Tiger attack helicopters	80	80
FELIN equipment	22,588	22,588
VBCI armored vehicles	630	630

Sources: Directorate General of Armaments, "Equipement" ("Equipment"), available from *www.defense.gouv.fr/ga/equipement*; and French Court of Auditors, *Le bilan à mi-parcours de la loi de programmation militare* (*Midterm Review on the Military Programming Law*), Paris, France, July 2012, p. 71, available from *www. livreblancdefenseetsecurite.gouv.fr/pdf/2012_07_11-cour-des_comptes_ rapport_thematique_bilan_lpm.pdf*.

Table 8-1. Orders for Major Weapons Systems.

STRENGTHS AND WEAKNESSES OF THE FRENCH ARMED FORCES

Strategic Forces.

Nuclear deterrence remains at the heart of French defense policy; it is seen as guaranteeing France a prominent place on the international stage and, as then–presidential candidate Hollande said in March 2012, protecting "the autonomy of our choices."[20]

France's nuclear forces consist of four ballistic missile–carrying submarines and a squadron of fighter bombers carrying cruise missiles. Ten percent of the overall defense budget and 20 percent of R&D funds go to maintaining these forces.[21] Although few in France question the need to retain a nuclear deterrent, some have argued that the aerial component is not required to sustain deterrence and, hence, could be shed to save money.[22] But as a recent report of the French Senate points out, the government is not facing an immediate need to spend large new sums to maintain its nuclear deterrent.[23] Previous investments in modernization have resulted in the deployment of a new generation of SSBNs; acquisition of a new ballistic missile and an advanced medium-range cruise missile; and the addition of the Rafale, a fourth-generation fighter jet, to its aerial nuclear strike force.

Army.

The French army retains 106,000 soldiers in 81 specialized regiments, making it one of the largest armies in Europe. It is also one of the best equipped with VBCIs, a FELIN, CAESER self-propelled howitzers, and Tiger attack helicopters.[24] But getting the most out of this equipment requires sustained training. In 2012, the army's days for training were down to 105, even though the law governing the French military for 2009–14 had authorized 120. The French court of auditors (Cour des Comptes) noted that even this level was somewhat misleading in that much of the training activity is focused on units deploying for low-intensity or counterinsurgency operations overseas, meaning the army has less time to hone other skills in areas of "high-intensity" conventional combat.[25]

Again, in an effort to reconcile reduced resources with the necessities of keeping the force modernized and trained, the white paper states that the army will make significant cuts to existing fleets of tanks and combat vehicles, while moving forward with a new generation of SCORPION networked armored combat vehicles. At the same time, however, the 2013 white paper has called for cutting in half the 2008 white paper's goal of being able to deploy 30,000 French troops. As the French chief of staff said, the 2008 white paper's objective was "unattainable," given current resources.[26] This reduction in capability has been criticized by others, including General Vincent Desportes, former head of the Joint Service Defense College, who suggested that this and other measures laid out in the 2013 white paper would "[relegate] France to a second tier. France will be unable to influence major strategic options internationally. Its role will be that of a junior partner."[27]

Although the French army is relatively small, the operational skills of its helicopter pilots are considerable, including the ability to conduct missions at night.[28] The recent experience in Mali, and, above all, in Afghanistan, has shown the importance of having a relatively large fleet of multirole helicopters available. According to the 2013 white paper, the goal is for the army to be equipped with 140 reconnaissance and attack helicopters, 115 tactical helicopters, and 30 tactical drones.[29]

Navy.

The French navy now has one aircraft carrier, 75 vessels and logistics ships, four nuclear-powered SSBNs, six nuclear-powered attack submarines, and less than 40,000 men. Since 2008, navy personnel have been cut by 6,000. Nineteen ships were taken out of service between 2009 and 2012, and only four new ships were added. According to Admiral Bernard Rogel, navy chief of staff, the size of the French fleet is "sufficient but just barely."[30] That said, the navy's budget for 2013 is set at €4.273 billion, the highest budget for equipment in the French armed forces.

Moreover, the French navy is one of the best trained in Europe, with 91 days at sea in 2010 and 92 in 2011. Major components of the fleet—SSBNs, amphibious assault ships, naval fighters, marine helicopters, and aircraft carriers—have recently been modernized or are in the process of being modernized. Plans are for France to replace its six Rubis-class, nuclear-powered attack submarines with the latest generation of Barracuda-class submarines, although only one is currently under construction. Finally, the 2013 white paper states that existing shortfalls in other parts of the fleet will be addressed with the acquisition of "15 first-class frigates, about 15 patrol vessels and six surveillance frigates, as well as maritime patrol aircraft and a mine warfare capability sufficient to protect our approaches and projection in expeditionary operations."[31]

Given budget constraints, however, there are concerns that orders for the FREMM frigates may still be cut back. France has already reduced its orders from 18 in 2005 to 11 in 2008. In June 2013, reports said the final number may be as low as nine or even eight.[32] But since the purpose of this program was to acquire frigates capable of performing missions that are currently carried out by several vessels, any reduction in the order will both increase the unit price for new FREMMs and require costly overhauls and modifications to existing platforms, such as the older La Fayette–class frigates.[33]

Air Force.

The French air force has also undergone profound changes since 2008. Personnel numbers dropped from 66,000 to 50,000, its air fleet was reduced by 30 percent, six fighter squadrons were disbanded, and eight air bases in France plus another four overseas were closed.

The 2013 white paper has announced that the air force fleet will be further reduced; the stated objective is 225 aircraft in place of the 300 planned in 2008. This means the air force will also reduce its orders for the Rafale multirole fighter and will look to extend the life of existing Mirages. In addition, the air force will be reducing the number of aircraft available for major operations from 70 to just 45.[34]

The air force is the branch of the armed forces with the most obvious capability gaps that, in turn, are in tension with France's efforts to maintain its strategic autonomy. The French fleet lacks long-distance strategic airlift. France has no equivalent to the U.S. Air Force's C-5 Galaxy or C-17 Globemaster III. France's fleet of smaller tactical transport aircraft is composed of 54 C-130 Hercules and Transall aircraft.[35] The lifespan of the C-160 Transall, which was put into service in 1967, has had to be extended because of delays in production and deliveries of Airbus A400Ms.

But with "downtimes" for repairs more frequent than newer planes, the C-160 has been expensive to maintain and operate. Further, the eight CASA/IPTN CN-235s acquired to fill the gap do not meet force projection needs, as was the case of the 2013 operation in Mali that required air logistic support from French allies.[36] In addition, resource constraints resulting from operations in Libya and Mali have impacted flight hours available to French pilots for training. The situation is particularly worrisome for transport pilots, who have had an activity level of only 287 hours instead of the planned 400.[37] Finally, it is worth noting that while the 2008 white paper set as an objective 70 tactical transport aircraft for the French air fleet, the 2013 paper lists "about 50."

The second significant gap in the air force's capabilities concerns tanker aircraft; in both the operation over Libya and the operation over Mali, the French required allied tanker support. In Operation UNIFIED PROTECTOR (Libya), for example, the United States performed about 70 percent of in-flight refueling missions, whereas France performed only about 10 percent.[38] The A330 MRTT is intended to replace the current aging fleet of French tankers. The first delivery of the plane, however, is not expected until 2017 at the earliest; the last is not expected until 2024.[39] The air force was planning to order 14 planes, but that number has now dropped to 12 tankers—a number that is probably insufficient if recent operations are a benchmark.[40]

Finally, the military intervention in Libya in 2011 also revealed the French air force's lagging capacity to neutralize land-based air-defense systems. In this instance, most Suppression of Enemy Air Defenses (SEAD) missions were performed by American forces despite the fact that Libya's air defenses were relatively weak.[41] France could potentially modify the armement air-sol modulaire (AASM, a modular air-to-ground missile) to carry a passive electromagnetic homing system to give it the SEAD capacity it currently lacks.[42] However, until it can acquire such a system, the air force will not have the ability to take the lead in similar air operations.

Intelligence.

The white paper says intelligence "must serve political and strategic decision-making as much as it serves planning and tactical conduct of operations. It should also shed light on our foreign and economic policies."[43] French intelligence services are known for their efficiency even though they have fewer resources with which to work than their major allies: 1.3 percent of the defense ministry's budget was designated for intelligence, or €655 million in appropriations, in 2013.[44]

Since the 2008 white paper, the government has placed increased emphasis on building up French intelligence capabilities, especially in the area of cyber, with special attention being paid to creating an offensive capability and in air- and space-based intelligence systems.[45] The equipment France uses to gather and analyze intelligence has changed significantly since the end of the Cold War. France now has strategic and tactical intelligence resources that it did not have during the wars in the Persian Gulf and Bosnia, when France was largely dependent on American strategic and tactical intelligence assets.

Maintaining and increasing that capability is also key to the French government's efforts to enhance France's strategic autonomy. While France was the coalition's second-largest contributor to intelligence, surveillance, and reconnaissance during operations in Libya in 2011, and despite the United States declaring it was "leading from behind," American Predator drones guided the French on their way to strike bunkers in Tripoli.[46] Accordingly, the French defense minister has affirmed that:

> several programs, too long delayed, have now been decided on and amplified: observation satellites, electronic listening satellites, embedded resources in airborne platforms, combat and tactical unmanned aerial vehicles (UAVs), and light surveillance and observation aircraft with their sensors.[47]

To that end, France's goal is also to have at its disposal 12 UAVs and seven detection and surveillance aircraft, in comparison with the four detection and surveillance aircraft in service today.[48]

Priority Zones and Pooling and Sharing.

For 20 years, the French military has been involved in numerous operations abroad, with the justification being, *inter alia*, the responsibility to protect the innocent, the war against terrorism, humanitarian crises, and missions of stability or peacekeeping. A review of French military interventions reveals that African conflicts are a French trademark. Since 1990, French armed forces have been involved in more than 20

African operations including in Rwanda, Somalia, Zaire, Comoros, Cameroon, the Republic of the Congo, Côte d'Ivoire, the Democratic Republic of the Congo, the Gulf of Aden, Chad, Libya, and Mali.

Several of these interventions have involved the commitment of significant French military resources. In Operation LICORNE in the Côte d'Ivoire, French troop presence increased to a height of 1,600. In Operation HARMATTAN in Libya, France committed fighter and reconnaissance aircraft, aerial refuelers, an airborne command and control plane, an aircraft carrier, an amphibious assault helicopter carrier, frigates, destroyers, and submarines. In Operation SERVAL in Mali, more than 4,000 French soldiers were deployed and more than 1,000 remain in country to support the new government and conduct stability operations. Even more recently, in December 2013, France sent 1,200 troops to the Central African Republic to help restore order and disarm the Muslim militias who had deposed the country's president earlier in the year.[49]

French forces have been deployed outside the African theater as well. Since the Cold War's end, French troops have been involved in several multilateral interventions, including the First Gulf War (contributing nearly 18,000 military personnel); the conflicts in the Balkans; in Lebanon as a major contributor to the United Nations Interim Force; and in Afghanistan, where France deployed more than 60,000 soldiers from 2001 to 2012.

But the cost of these interventions combined with the apparent lack of success in missions as in the case of Afghanistan have resulted in France's decision to scale back its strategic sights and more strictly define "priority zones" for its military interventions. These priority zones are the European periphery, the Mediterranean Basin, the Persian Gulf, the Indian Ocean, and Northern Africa from the Sahel to the equatorial countries. The Sahel corresponds to the zone of vital interest that France should, it believes, be able to defend. As a result of France's historical presence, Africa is home to one of the largest groups of French expatriates: more than 210,000 French citizens live there.[50] Additionally, special defense agreements with Gabon, Senegal, Djibouti, and Chad give France a higher degree of legitimacy and an operational advantage when it comes to intervening in the region.

Moreover, major security challenges exist just outside the gates of Europe, with the rise in terrorism and criminal activities resulting from instability in the wake of the Arab Spring and the need to secure major resource and supply routes from the Middle East, Africa, and South Asia. Indeed, America's planned pivot to Asia and its reluctance to intervene further in the Middle East was duly noted in the 2013 white paper:

> The evolving strategic context may place our country in a position in which we are obliged to take the initiative in operations, or to assume, more often than in the past, a significant part of the responsibilities involved in conducting military operations.[51]

Given this strategic context, it is no surprise that France is attempting once again to jumpstart the European common defense effort. After the principle of differentiation within French forces and the concept of strategic autonomy, the white paper's third pillar of French defense policy is greater reliance on the pooling and sharing of defense capabilities by European powers. The decrease of European defense capabilities combined with the budgetary crisis is seen as an opportunity to promote greater cooperation among countries in defense of European vital interests. As the white paper puts it, France aims for greater pooling of capabilities on the European level to "replace forced dependency with organized inter-dependency."[52] To obtain this goal, however, Europe's capitals will need to establish a deeper consensus on the most important security issues they might face and a greater willingness to address them by joint action.

It will also require a tough-love approach to Europe's national defense companies who, with the continuing decline in European defense spending, face a smaller market at home and increased competition from the United States, Russia, and China abroad. The risk is that the budgetary pressures on investment in the short term will translate into a general decline in the specialized industrial know-how that the companies must maintain if they are to remain competitive. To avoid this, European capitals will have to put aside the desire to protect their respective national companies and allow a continent-wide restructuring of Europe's defense industry to move forward.[53]

CONCLUSION

French military ambitions are increasingly limited by the economic crisis and France's fiscal problems. As a percentage of French GDP, defense is less of a national priority today. (In 1997, the military budget equaled 2 percent of GDP; today, it stands at approximately 1.5 percent.) That said, France's decision to intervene in Mali this past year is a sober reminder of France's need to maintain serious military capabilities to protect its interests and address the existing gaps in needed capabilities.

But the actual risk France runs lies less in the condition of today's French forces than in their future state. Essentially freezing the defense budget for several years as planned will cost the French military in a number of ways. By not replacing equipment in an orderly fashion, an increasing portion of the defense budget will go to maintaining aging equipment; already, the amount devoted to maintenance is up by 8 percent in 2013.[54] Indeed, according to the French chief of the defense staff, estimates in 2013 for the availability of armored personnel carriers, frigates, and combat planes would be 40, 48, and 60 percent, respectively.[55] While French forces are no longer in Afghanistan, budget constraints will make it more difficult to keep training levels up to previous standards, which is a must for some units such as joint tactical battalions that, moving forward, will be the core building block for French interventionist forces. Moreover, if France wants to continue to be a global leader in developing and fielding military technologies, it will need to maintain a significant level of investments in R&D. In fact, before the expiration of the recently passed military programming law in 2019, France will need to have begun work on next-generation weapons systems if it expects to sustain itself as a modern fighting force.

Naturally consumed with dealing with today's problems, the fact remains that it is President Hollande's responsibility to plan for the armed forces of 2035. Even though Defense Minister Jean-Yves Le Drian has stated that he intends to safeguard the defense budget until 2016, French Parliament members' temptation to make defense even more so the "bill payer" for reducing the government's deficit will remain. Past history provides little support for that pledge as no multiyear military programming law passed by the legislature has ever escaped modification by the government and French legislators in subsequent years. The state of the French military is at a critical juncture. A wrong step now could leave France with a future military that can no longer adequately address the country's security interests or sustain its goal of strategic autonomy.

ENDNOTES - CHAPTER 8

1. This chapter was originally published as an essay on February 4, 2014.

2. Government of France, *Livre Blanc: Défense et Sécurité Nationale 2013* (*White Paper: Defense and National Security 2013*), Paris, France, 2013, available from *www.livreblancdefenseetsecurite.gouv.fr/pdf/le_livre_blanc_de_la_defense_2013.pdf.* For an example of press reports about expected cuts to defense, see Michel Cabirol, "Un tsunami s'annonce sur le budget de la Défense" ("Promise of a Tsunami in the Defense Budget"), *La Tribune*, July 12, 2012, available from *www.latribune.fr/entreprises-finance/industrie/aeronautique-defense/20120711trib000708598/un-tsunami-s-annonce-sur-le-budget-de-la-defense.html.*

3. Heads of the seven major French defense firms, Thales, Nexter, DCNS SA (Dassault Aviation, Safran, MBDA, and European Aeronautic Defence and Space Company NV) sent a letter to the French president stating, "It is essential that industrial and socio-economic issues be taken into account as seriously as budget issues." In addition, five political groups from across the political spectrum stated their concerns about reductions in defense spending in July 2012 during a meeting of the French Senate's foreign affairs, defense, and armed forces committee. See "Inquiets, les industriels de l'armement demandent audience à l'Elysée" ("Concerned, the Arms Manufacturers Require Hearing at the Elysée"), *Les Échos*, March 13, 2013, available from *www.lesechos.fr/13/03/2013/LesEchos/21395-064-ECH_inquiets–les-industriels-de-l-armement-demandent-audience-a-l-elysee.htm*; and French Senate, "Le seuil de 1.5% du PIB consacré à l'effort de défense est incompressible" ("The 1.5 Percent of GDP Devoted to Defense is an Incompressible Amount"), July 26, 2012, available from *www.senat.fr/presse/cp20120726a.html.*

4. Government of France, "LOI n° 2013-1168 du 18 décembre 2013 relative à la programmation militaire pour les années 2014 à 2019 et portant diverses dispositions concernant la défense et la sécurité nationale" ("Law No. 2013-1168 of December 18, 2013, Pertaining to the Military Program for the Years 2014 to 2019"), January 1, 2014, available from *www.legifrance.gouv.fr/affichTexte. do;jsessionid=?cidTexte=JORFTEXT000028338825.*

5. Entretien du président de la République sur France 2" ("Interview with the President of the Republic on France 2"), *France 2,* March 28, 2013, available from *www.elysee.fr/interviews/article/ entretien-du-president-de-la-republique-sur-france/.*

6. Vincent Lamigeon, "Défense et livre blanc: la demi-victoire budgétaire des militaires et de l'industrie de l'armement" ("Defense White Paper: Fiscal Half-Victory for the Military and the Defense Industry"), *Challenges,* April 4, 2013, available from *www. challenges.fr/economie/20130404.CHA7981/bercy-defense-et-livre- blanc-la-demi-victoire-budgetaire-des-militaires-et-de-l-industrie-de-l- armement.html.*

7. See Government of France, "LOI no 96-589 du 2 juillet 1996 relative à la programmation militaire pour les années 1997 à 2002" ("Law No. 96-589 of July 2, 1996, Related to the the Military Programming Law for the Years 1997 to 2002"), July 2, 1996, available from *legifrance.gouv.fr/affichTexte.do;jsessionid=9CB2BFEB3123BFA 5C927AEA3F57F2FE2.tpdjo09v_1?cidTexte=JORFTEXT0000005602 00&categorieLien=id.*

8. International Institute for Strategic Studies, "NATO and Non-NATO Europe," *The Military Balance 1998,* Vol. 98, No. 1, p. 50; International Institute for Strategic Studies, "Europe," *The Military Balance 2003,* Vol. 103, No. 1, p. 39.

9. International Institute for Strategic Studies, "Chapter Four: Europe," *The Military Balance 2013,* Vol. 113, No. 1, p. 131.

10. Emmanuel Cugny, "Budget Defense 2014–2019: Les longs soupirs de la Grande mette" ("Defense Budget 2014–19, The Army's Big Sigh"), *France Info,* November 26, 2013, available from *www.franceinfo.fr/economie/tout-info-tout-eco/budget-defense- 2014-2019-rendez-vous-a-l-assemblee-nationale-1227037-2013-11-26.*

11. During 1997–2002, €12.6 billion allotted to the procurement budget was not distributed.

12. See Government of France, *Livre Blanc: Défense et Sécurité Nationale 2008* (*White Paper: Defense and National Security 2008*), Paris, France, 2008, available from *archives.livreblancdefenseetsecurite.gouv.fr/2008/information/les_dossiers_actualites_19/livre_blanc_sur_defense_875/livre_blanc_1337/livre_blanc_1340/index.html*.

13. French Court of Auditors, *Le bilan à mi-parcours de la loi de programmation militaire* (*Midterm Review on the Military Programming Law*), Paris, France, July 2012, p. 71, available from *www.livreblancdefenseetsecurite.gouv.fr/pdf/2012_07_11-cour_des_comptes_rapport_thematique_bilan_lpm.pdf*.

14. Véronique Guillermard and Cécile Crouzel, "Défense: 5,5 milliards de report de commandes" ("Defence: 5.5 Billion in Deferred Orders"), *Le Figaro*, September 28, 2012, available from *www.lefigaro.fr/conjoncture/2012/09/28/20002-20120928ART-FIG00443-defense-55-milliards-de-report-de-commandes.php*.

15. *Livre Blanc: Défense et Sécurité Nationale 2013*, p. 69.

16. Vigipirate is France's national security alert system.

17. Emmanuel Grasland and Alain Ruello, "Les industriels de l'armement adressent un message d'alerte à Hollande" ("Defense Industry Players Send Hollande a Word of Warning"), *Les Échos*, March 13, 2013, available from *www.lesechos.fr/13/03/2013/lesechos.fr/0202639098086_les-industriels-de-l-armement-adressent-un-message-d-alerte-a-hollande.htm*.

18. *Direction Générale de l'Armement* (*General Directorate for Armament,* or DGA) is the French government defense procurement agency.

19. See French Senate, *Rapport D'Information N°680* (*Information Report No. 680*), July 18, 2012, p. 42, available from *www.livreblancdefenseetsecurite.gouv.fr/pdf/2012_07_18-rapport_senat-format_emploi_armees.pdf*.

20. "Discours de François Hollande sur la défense nationale à Paris le 11 mars 2012" ("François Hollande's Speech on National Defense in Paris"), March 11, 2012, available from *www.gilbert-roger.fr/files/fh-de%25CC%2581fense-11-mars.pdf*.

21. Robert Carmona, "Un budget de transition pour la Défense en 2013" ("A Transition Budget for Defense in 2013"), *La Tribune*, No. 313, March 2013.

22. "Hervé Morin veut supprimer la composante aérienne de la dissuasion pour faire des economies" ("Hervé Morin Wants to Remove the Air Component of Deterrence to Save Money"), *Zone Militaire*, January 10, 2013, available from *www.opex360. com/2013/01/10/herve-morin-veut-supprimer-la-composante-aerienne-de-la-dissuasion-pour-faire-des-economies/*.

23. See French Senate, *Rapport d'Information N°668* (*Information Report No. 668*), July 12, 2012, available from *www.senat.fr/rap/ r11-668/r11-6681.pdf*.

24. The 110e Régiment d'Infanterie (110th Infantry Regiment) will be dissolved in 2014.

25. French Court of Auditors, *Le bilan à mi-parcours de la loi de programmation militaire*, pp. 65-66.

26. Michel Cabirol, "La France, l'adieu aux armes?" ("France, A Farewell to Arms?"), *La Tribune*, August 26, 2012, available from *www.latribune.fr/entreprises-finance/industrie/aeronautique-de-fense/20120826 trib000716279/la-france-l-adieu-aux-armes-.html*.

27. "Général Vincent Desportes: La défense française est dégradée et déséquilibrée" ("General Vincent Desportes: The French Defense Is Degraded and Unbalanced"), *La Croix*, May 1, 2013, available from *www.la-croix.com/Actualite/France/General-Vincent-Desportes-La-defense-francaise-est-degradee-et-desequili-bree-2013-05-01- 955039*.

28. Jean-Dominique Merchet "Aérocombat: Nos alliés sont bluffés!" ("Air Combat: Our Allies Are Bluffed!"), *Marianne*, February 1, 2013, available from *www.marianne.net/blogsecretdefense/ Aerocombat-Nos-allies-sont-bluffes-_a932.html*.

29. Government of France, *Livre Blanc: Défense et Sécurité Nationale 2013*, p. 73.

30. See French National Assembly Committee on National Defense and Armed Forces, "Audition de l'admiral Bernard Rogel, chef d' état -major de la marine" ("Testimony of General Bernard Rogel, Chief of Staff of the Navy"), July 18, 2012, available from *www.assemblee-nationale.fr/14/cr-cdef/11-12/c1112006.asp*.

31. Government of France, *Livre Blanc: Défense et Sécurité Nationale 2013*, p. 110.

32. Jean-Marc Tanguy, "Le Livre blanc lave plus blanc que blanc" ("The White Paper Washes Whiter than White"), *Le mamouth* blog, April 30, 2013, available from *lemamouth.blogspot.com/2013/04/le-livre-blanc-lave-plus-blanc-que-blanc.html*; and Jean-Dominique Merchet, "Marine: ce sera 8 Fremm au lieu de 11" ("Navy: That Will Be 8 FREMM Instead of 11"), *Marianne*, April 30, 2013, available from *www.marianne.net/blogsecretdefense/Marine-ce-sera-8-Fremm-au-lieu-de-11_a1028.html*.

33. The decision on whether to complete the FREMM program or begin work on an alternative program will be made by 2016. See "Le remplacement des trois dernières FREMM par des FTI n'est plus tabou" ("The Replacement of Three Last FREMM by FTI Is No Longer a Taboo"), *Mer et Marine*, October 10, 2013, available from *www.meretmarine.com/fr/content/le-remplacement-des-trois-dernieres-fremm-par-des-fti-nest-plus-tabou*.

34. "Projet de LPM: Un moindre mal" ("MPL Project: A Lesser Evil"), *Mer et Marine*, August 5, 2013, available from *www.meretmarine.com/fr/content/projet-de-lpm-un-moindre-mal*.

35. Figures refer to 2012.

36. "Qui participe à l'opération Serval au Mali?" ("Who Is Participating in the Operation Serval in Mali?"), *Le Monde*, January 29, 2013, available from *www.lemonde.fr/afrique/article/2013/01/29/qui-participe-a-l-operation-serval-au-mali_1824111_3212.html*.

37. See French Senate, *Rapport d'Information N°668*, p. 42.

38. See French National Assembly Committee on National Defense and Armed Forces, *Compte rendu n°18* (*Report No. 18*), October 30, 2012, available from *www.assemblee-nationale.fr/14/cr-cdef/12-13/c1213018.asp#P5_424*.

39. French Court of Auditors, *Le bilan à mi-parcours de la loi de programmation militaire*, p. 61.

40. Government of France, *Livre Blanc: Défense et Sécurité Nationale 2013*, p. 74.

41. See French National Assembly Committee on National Defense and Armed Forces, *Compte rendu n°18*.

42. Lieutenant Yohan Droit, "Mission et capacités SEAD — Une perspective de l'armée de l'air" ("SEAD Mission and Capabilities — A View of the Air Force"), Paris, France: Centre d'études stratégiques aérospatiales, September 2012, available from *www.cesa.air.defense.gouv.fr/IMG/pdf/Mission_et_capacites_SEAD-_sept_2012.pdf*.

43. Government of France, *Livre Blanc: Défense et Sécurité Nationale 2013*, p. 70.

44. See French Senate, *Projet de Loi N°23* ("Law Draft No. 23"), October 21, 2013, available from *www.senat.fr/leg/tas13-023.html*; and French National Assembly Committee on National Defense and Armed Forces, *Compte rendu n°56* (*Report No. 56*), February 20, 2013, available from *www.assemblee-nationale.fr/14/cr-cdef/12-13/c1213056.asp*.

45. Johan Galtung, "Les formes alternatives de défense: l'exemple européen" ("Alternative Forms of Defense: The European Example"), *Érudit*, Vol. 20, No. 3, 1989, pp. 625–645, available from *www.erudit.org/revue/ei/1989/v20/n3/702545ar.html*.

46. Catherine Durandin, *Le déclin de l'armée française* (*The Decline of the French Military*), Paris, France: François Bourin, 2013, p. 84.

47. Jean-Yves Le Drian, *Discours aux commandeurs* (*Speech to the Commanders*), Paris, France: Ministry of Defense, École Mili-

taire, April 29, 2013, available from *www.defense.gouv.fr/content/download/206259/2287506/file/DISCOURS-DU-MINISTRE-AUX-COMMANDEURS-LE-LUNDI-29-AVRIL-2013-A-L-ECOLE-MILITAIRE.pdf.*

48. France acquired two MQ-9 Reapers last year. See Pierre Tran, "French AF Conducts First Reaper Flight," *Defense News*, January 16, 2014, available from *www.defensenews.com/article/20140116/DEFREG01/301160030/French-AF-Conducts-First-Reaper-Flight.*

49. Cordélia Bonal, "Centrafrique: pourquoi la France intervient" ("Central Africa: Why France Is Intervening"), *Libération*, December 5, 2013, available from *www.liberation.fr/monde/2013/12/05/pourquoi-la-france-va-en-centrafrique_964366.*

50. "French Overseas," Paris, France: French Ministry of Foreign Affairs, December 31, 2012, available from *www.diplomatie.gouv.fr/en/french-overseas/.*

51. Government of France, *Livre Blanc: Défense et Sécurité Nationale 2013*, p. 63.

52. *Ibid.*, p. 5.

53. The proposed merger of European Aeronautics and Defense and Space (EADS) and Britain's BAE Systems is a good example of what probably should have taken place but did not. For analysis of why the proposed merger did not work out, see Carol Matlack, "Why the EADS-BAE Deal Collapsed," *Bloomberg Businessweek*, October 10, 2012, available from *www.businessweek.com/articles/2012-10-10/why-the-eads-bae-deal-collapsed.*

54. See French Senate, *Projet de Loi N°23.*

55. Frédéric Charillon, "France's New Military Budget: Rethinking Power," *RUSI*, November 13, 2013, available from *www.rusi.org/publications/newsbrief/ref:A528361D338FD6/#.UuZYVRAo5pg.*

CHAPTER 9

TAIWANESE HARD POWER:
BETWEEN A ROC AND A HARD PLACE[1]

Michael Mazza

KEY POINTS

- Over the past decade, the cross–Taiwan Strait military balance has shifted in favor of the Chinese military.
- Taiwan's efforts to meet that challenge have been slowed by insufficient defense budgets; difficulties in establishing an all-volunteer, active duty force; and a complicated political and economic relationship with the mainland.
- The United States has a statutory interest in and obligation to help Taiwan maintain an adequate defense posture, but in recent years has fallen short of meeting those goals.

Taiwan's 2013 *National Defense Report*, a biennial publication of Taiwan's Ministry of National Defense (MND), paints a bleak picture of the island's future security. It asserts that China "plans to build comprehensive capabilities for using military force against Taiwan by 2020."[2] The People's Liberation Army (PLA) is apparently well on its way to achieving that objective.

The report describes worrisome advances across the spectrum of PLA capabilities. According to the MND, the PLA's intelligence, surveillance, and reconnaissance (ISR) capabilities are sufficient "to support the use of military force for resolving the Taiwan issue." The ground force can already conduct a landing on, and seizure of, Taiwan's offshore islands, while the Chinese navy can "effectively blockade the Taiwan Strait and seize near shore islands" and "blockade key air space." The air force, for its part, is currently "capable of fighting for air superiority and control over the first island chain," which stretches from the Japanese home islands through Taiwan and south to the Philippines.[3]

The report also highlights advances in China's missile force, notably, the fielding of advanced anti-ship ballistic missiles and the deployment of more than 1,400 missiles with conventional warheads aimed against Taiwan. The bottom line: "Combined with the Navy and Air Force, the PLA is now capable of conducting large scale joint firepower strikes and denying foreign forces from intervening in disputes across the Taiwan Strait."[4]

Together, these developments mark a major shift in the balance of power in the Taiwan Strait. Indeed, not all that long ago, the balance favored Taiwan. For example, in 2000, Michael O'Hanlon argued that "China cannot invade Taiwan, even under its most favorable assumptions about how a conflict would unfold."[5] According to O'Hanlon, even coercive operations short of a full-scale invasion would have been difficult for the PLA to pull off.

In the MND's previous *National Defense Report*, released in 2011, negative trends were evident but not so starkly stated. Now the MND assesses that the PLA is only 6 years away from fielding an effective invasion force, and that it can already prevent outside powers from intervening in a timely way. How did Taiwan arrive at this juncture?

DEFENSE BUDGET TRENDS

While Beijing has sustained 2 decades of double-digit growth in its defense budget, Taipei has not evinced a similar commitment to defense spending. In 1996, the year of Taiwan's first free presidential election, Taiwan's military expenditures stood at U.S.$12.9 billion (in constant 2011 dollars), accounting for 4.1 percent of gross domestic product (GDP). The defense budget's share of GDP had already been trending downward, but that trend accelerated from 1996 onward. Between 1991 and 1995, the average percentage change in the military's share of GDP was -4.4 percent; between 1996 and 2000, that rate dropped to -8.3 percent.[6]

Today, Taiwan only commits 2 percent of its GDP to defense, well short of the 3 percent goal set by both previous Taiwanese Presidents Chen Shui-bian and his successor, Ma Ying-jeou.[7] In 2012, Taiwan spent U.S.$10.5 billion on defense, 20 percent less than it was spending in 1996 (again, in constant 2011 dollars). (See Figure 9-1.)[8]

Defense spending as a share of GDP provides a rough measure of a country's overall commitment to its defense. The military budget's share of the total national budget provides a similar indicator and points to how the government prioritizes defense spending in any given year. A Congressional Research Service analysis found that Taiwan's military budget was responsible for 22.8 percent of total government spending in 1996. In 2013, that share stood at only 16.4 percent.[9]

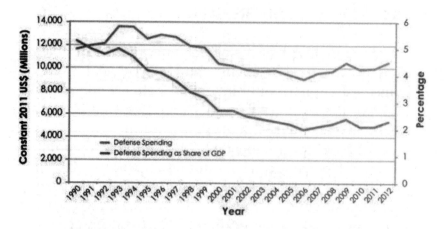

Source: Stockholm International Peace Research Institute (SIPRI) Military Expenditure Database, available from *www.sipri.org/ research/armaments/milex/milex_database*.

Figure 9-1. Taiwan's Defense Spending, 1990-2012.

TAIWAN'S SECURITY ENVIRONMENT

These trends are surprising when one considers the inhospitable security environment in which Taiwan has found itself in the past 2 decades. In 1995, the PLA conducted a series of missile tests in waters around Taiwan to express its displeasure at former Republic of China (ROC) President Lee Teng-hui's visit to the United States. China did so again in the lead-up to Taiwan's 1996 presidential election. In both cases, the United States responded by sending aircraft carriers to the region.

In 2005, Washington was seemingly preoccupied with the wars in Iraq and Afghanistan and Beijing was facing a Taiwanese president (Chen Shui-bian) who prioritized asserting Taiwan's status as an independent democratic state. Against this backdrop, Beijing promulgated the Anti-Secession Law "for the purpose of opposing and checking Taiwan's secession from China by secessionists in the name of 'Taiwan independence'." Although the law asserts a preference for pursuing "peaceful unification," Article 8 stipulates China's right to use "non-peaceful means and other necessary measures to protect China's sovereignty and territorial integrity."[10]

Kuomintang candidate Ma Ying-jeou's election as Taiwan's president in 2008 and his subsequent cross-Strait economic policies helped stabilize relations between Taipei and Beijing. Yet even as Beijing acted with less overt hostility toward Taiwan, it pursued policies that did little to reduce the ROC's international isolation, continued to increase its military capabilities vis-à-vis Taiwan, and left the island generally less secure. In 2011, two PLA fighter aircraft even entered Taiwanese airspace in an attempt to scare off an American spy plane, a reminder that, to the leaders in Beijing, Taiwan has no sovereign airspace.

China's attempt to wrest control of the disputed Senkaku Islands, which Taiwan also claims, from Japan threatens to turn Taiwan's northern flank. Beijing's apparent aim of turning the South China Sea into a Chinese lake likewise threatens Taiwan's security. Chinese success there would not only have implications for Taipei's own claims in the sea but would also enhance Beijing's ability to coerce Taiwan militarily.

China's East China Sea Air Defense Identification Zone (ADIZ) and its behavior in the South China Sea, moreover, amount to an outright challenge to freedom of navigation through the seas and skies, on which Taiwan depends for its economic vitality. In short, Taiwan's security environment has deteriorated substantially since the Taiwan Strait crisis of 1995-96.

POLITICAL ROADBLOCKS

With Taiwan's transition to full-fledged democracy in the mid-1990s, Taipei began finding it difficult to sustain previous levels of defense spending. As in many democratic societies, the influence of various interest groups has increased over time. Over the last 2 decades in particular, the Taiwanese government has dedicated relatively larger shares of the national budget toward social welfare, economic development, education, and pension payments, putting downward pressure on defense spending.[11]

Moreover, with Taiwan's anemic birthrate, the traditional welfare system, in which children take care of their parents, and siblings assist each other in times of need, has shown signs of fraying. As a result, in recent years the government has eased criteria for securing access to welfare, which has once again lessened the government revenues available for national security.[12]

Even when the leadership has wanted to spend more on defense, domestic politics have gotten in the way. Chen Shui-bian, who served as ROC president from 2000 to 2008, was in favor of more social spending and a strong defense. In 2001, the George W. Bush administration approved a major arms package for Taiwan, which included submarines, antisubmarine warfare aircraft, torpedoes, and anti-ship cruise missiles, among other systems.

But in the Legislative Yuan (Taiwan's legislature), the national security debate became highly politicized. With the legislature failing to approve of spending funds that the executive branch had earmarked for these purposes, the defense budget actually declined during the Chen administration.

The Ma Ying-jeou administration likewise has had difficulty boosting the defense spending level to its stated goal of 3 percent of GDP. This has been in part because of the global economic downturn and its impact on the export-dependent Taiwanese economy.

But Taiwan's struggling economy does not alone explain the administration's failure to reach its defense spending target. Although President Ma has consistently claimed a need for Taiwan to maintain a strong defense, his cross-Strait policies may also make it difficult to sustain public support for more defense spending.

Under Ma, Taipei has succeeded in reducing tensions across the Taiwan Strait. The president's "three nos" policy—no unification, no independence, no use of force—has served to reassure Beijing after the 8-year, independence-minded Chen Shui-bian administration. Ma's cross-Strait policy has moreover emphasized opportunities for cooperation with the mainland. The signature achievement of this policy is the Economic Cooperation Framework Agreement, which loosened restrictions on cross-Strait trade and paved the way for Taiwan to complete free-trade agreements with Singapore and New Zealand.

Ma, of course, has been eager to tout the successes of his policies. But a side effect has been for his administration to underemphasize those aspects of Beijing's policies that continue to threaten Taiwan. To emphasize those aspects would be to undercut, at least rhetorically, the Ma administration's claimed accomplishments. This perhaps explains why Taipei's reaction to China's 2013 ADIZ announcement, while critical, was more muted than that of its Japanese and South Korean neighbors.

President Ma has argued that a robust defense allows Taiwan to deal effectively with China from a position of strength. This is sensible. But in telling the Taiwanese public that all is going swimmingly in cross-Strait relations, he may well be weakening his own calls for a strong military deterrent.

TAIWAN'S DEFENSE AND MILITARY STRATEGIES

While the island's level of defense spending may fall short of stated goals, the military has nevertheless added new capabilities in recent years. Since 2001, Taiwan has purchased Kidd-class destroyers, P-3C Orion maritime patrol aircraft, Patriot Advanced Capability-3 (PAC-3) defense missiles, Black Hawk utility helicopters, Osprey-class coastal mine-hunter ships, Apache attack helicopters, retrofits of its 145 F-16 fighter jets, and numerous sea, ground, and air-launched munitions, all from the United States.[13] Domestically, Taiwan has been upgrading its Indigenous Defense Fighters (IDFs), building fast-attack missile boats, and developing anti-ship and land-attack cruise missiles.

Taiwan's second and most recent *Quadrennial Defense Review* (QDR), released in 2013, lays out strategies that are both consistent with previous planning documents and mindful of current conditions. Taiwan's national defense strategy rests on five pillars: war prevention, homeland defense, contingency response, conflict avoidance, and regional stability. The strategy equally emphasizes measures aimed at ensuring the Taiwanese military's ability to fight and those designed to ensure a fight will not be necessary.

For example, the MND claims it will "develop defense technologies, continue to procure defensive weapons, establish 'innovative and asymmetric' capabilities, and strengthen force preservation and infrastructure protection capabilities."[14] However, the ministry also vows to institute "information transparency measures" to "help enhance surrounding countries' understanding of the ROC's defense policy, objectives of military preparation and readiness, and contents of military activities," the aim being to "reduce distrust, miscalculation and misunderstanding."[15] The defense strategy also emphasizes collaborative approaches to security: promoting enhanced security dialogues and exchanges, working with others to establish regional "security mechanisms," and establishing programs to "jointly safeguard regional maritime and air security."[16]

Taipei's military strategy is more narrowly focused on how ROC forces will defend against threats to Taiwan's sovereignty and territorial integrity. However, the strategy's overarching theme—"resolute defense, credible deterrence"—is not well defined. MND's 2013 *National Defense Report* describes "resolute defense" thusly:

> A defense force that is only used when attacked by the enemy, and is the minimal force required only for defense. The defense force is also limited to protecting territorial integrity, and thus adopts a passive defense strategy.[17]

The 2013 QDR's description is somewhat more specific, describing a requirement "to be able to conduct fortified defense, reinforce and support, and recapture operations," but the emphasis remains defensive in nature.[18]

Here, there may be tension with the "credible deterrence" aspect of Taiwan's military strategy. As with "resolute defense," the QDR's definition of "credible deterrence" is problematic:

> The ROC Armed Forces should continue force training and combat preparation, effectively integrate the interoperability of weapon systems, enhance joint operational performance, and exert overall warfighting capabilities, forcing the enemy to consider the costs and risks of war, thereby deterring any hostile intention to launch an invasion.[19]

First, it is unclear what Taiwan's military intends to hold at risk that would effectively deter China. In contemplating an invasion, Beijing can be expected as a matter of course to consider the costs and risks of war. The QDR fails to explain how Taiwan will raise those costs and deepen those risks. Second, the military presumably wants to deter not only a full-scale invasion but also a missile barrage, blockade, or other coercive use of force. The omission is curious.

Looking beyond the "resolute defense, credible deterrence" slogan, however, the QDR offers more concrete plans for contending with the PLA. In particular, the QDR describes requirements to counter a blockade of the sea or air lines of communication, for joint interdiction of forces approaching from mainland China, and for ground forces capable of denying Chinese forces from establishing a beachhead.

The QDR goes on to emphasize the need for the continued development of joint warfighting capabilities based on the military's "innovative and asymmetric" concept, which recognizes that Taiwan cannot compete with China on quantitative grounds and should develop capabilities to target China's weaknesses.

Even so, the QDR describes a force that can contend with PLA forces in the air, at sea, on the ground, and in the cyber and electronic domains.

Enabling all activities in the future will be effective joint command, control, communication, computers, and ISR. According to the QDR, the military "must strengthen battlespace management, command, control, intelligence and early warning capabilities to accurately monitor enemy activities and flexibly execute force maneuver."[20]

What Taiwan requires, in short, is a highly skilled, innovative, high-tech force. But it is questionable whether Taiwan can successfully create this defense force, given resource and manpower constraints and shortcomings in U.S.-Taiwan defense cooperation.

SHIFTING FORCE STRUCTURE

Over the past 15 years, Taiwan has been shifting to a smaller, more high-tech force. In 1999, its armed forces consisted of 370,000 active duty members; that number dropped to 290,000 in the first half of the last decade. While Taiwan's military has pursued modernization across all three services, the largest force-structure changes have occured in the navy.

On the whole, the fleet has shrunk, both in numbers of ships and in average ship size. At the beginning of the century, Taiwan's navy had a traditional surface-warfare emphasis. The fleet included 12 destroyers, which constituted more than a third of Taiwan's principal surface combatants.

Since that time, the navy has retired all of those ships, replacing them with just four Keelung-class destroyers. The Keelungs, former American Kidd-class destroyers, are the largest warships the ROC Navy has ever operated. Equipped with SM-2 Block IIIA and RGM-84L Block II Harpoon missiles, the ships provided the navy with enhanced modern air defense and anti-surface warfare capabilities.

The navy has also focused on recapitalizing its fleet of small missile boats. Since 2005, it has retired all 48 of its 1970s-era Hai Ou-class ships, replacing them with 31 Kwang Hua VI-class vessels. The larger but stealthier Kwang Huas carry a slightly larger, more modern armament of anti-ship cruise missiles. In March 2014, the navy received the Tuo River, the first of up to 12 new fast-attack missile boats that have been dubbed "carrier killers" in Taiwan. Also described as corvettes, the craft will be stealthy and armed with eight anti-ship cruise missiles.[21]

Taiwan's navy continues to prioritize modernizing its undersea force as well. It has two Dutch submarines built in the 1980s and two World War II-era, U.S. GUPPY-class submarines, all four of which are in need of replacement. (The GUPPY-class submarines are now used only for training.)

In 2000, U.S. President George W. Bush agreed to sell Taiwan eight new diesel-electric submarines. For a variety of reasons, none of them have been built or sold. While the Taiwanese navy continues to insist that acquiring new submarines is a priority — they would be particularly useful for counterblockade and anti-surface warfare missions — it has continued to upgrade munitions for its two deployable Dutch boats. Last year, the navy received 32 submarine-launched anti-ship cruise missiles from the United States, which may also allow for strikes on Chinese coastal targets.[22]

The navy has been upgrading not only its fleet but also its maritime air capabilities. Most notable in this regard is the purchase of 12 P-3C Orion patrol aircraft from the United States. The navy received the first four of these planes in 2013; five more are set for delivery in 2014, with the remainder arriving in Taiwan by the end of 2015. With China enhancing its undersea force, the Orions will provide the ROC Navy with a proven anti-submarine warfare capability.

The military has likewise pursued army aviation upgrades, though new capabilities have only just begun to enter the force. In November 2013, Taiwan received the first batch of an expected 30 total AH-65E Apache Guardian attack helicopters from the United States, becoming the first foreign operator of the updated chopper. The helicopters, which will be a marked improvement over Taiwan's current AH-1W Cobras, will enhance Taiwan's ability to counter a cross-Strait invasion force and to prevent an enemy from establishing a beachhead.

The first delivery of UH-60M Black Hawk utility helicopters occurred in 2014; Washington approved the sale of a total of 60 Black Hawks to Taipei. These aircraft, which replace 1950s-designed UH-1H choppers, provide the army with greater mobility. They will be key assets for an army expected to play a greater role in responding to natural disasters—to which Taiwan is prone—and will provide the ability to move quickly around the mountainous island in the event of Chinese aggression.

One of the army's most important acquisitions in recent years has been the PAC-3 ground-based missile defense system. Taiwan currently operates one battery on the northern end of the island and has plans to add three more to the south.[23] PAC-3 missiles provide defense against cruise and ballistic missiles and enemy aircraft.

Although the army and navy have been successfully upgrading their aerospace capabilities, the air force has had difficulty doing so. The air force has been able to secure from the United States munitions and upgrades for its current fleet of F-16A/B fighter jets, but those upgrades are now at risk.

In its 2015 budget request, the U.S. Air Force deleted funding for the Combat Avionics Programmed Extension Suite, which was meant to "replace the avionics and radars for 300 U.S. F-16s" and for Taiwan's 146 F-16A/Bs.[24] Upgrades will remain available for Taiwan's planes, but likely at an additional cost of tens of millions of dollars. Whether those estimates grow and whether the Legislative Yuan will approve the additional expenditure remain open questions.[25]

Just as troubling, Taiwan has been unsuccessful in securing new aircraft needed to replace old F-5s and Dassault Mirage 2000s that must be retired. As a result, Taiwan will continue flying legacy aircraft at least over the next decade. Without upgrades to those aircraft, a cross-Strait air-power capability gap will continually grow in China's favor (see Figure 9-2).

Taiwan's Missile Program.

Taiwan has an active indigenous cruise missile program. While U.S. officials have at times expressed unease with the program, Taiwan has been undeterred in producing weapons it believes are necessary for the island's defense. In recent years, Taiwan has fielded two new cruise missiles: the Hsiung Feng IIE (HF-2E) and the Hsiung Feng III (HF-3).

The less controversial of the two is the HF-3, an anti-ship cruise missile that can be fired from land-based and seaborne platforms. Many of the navy's ships are

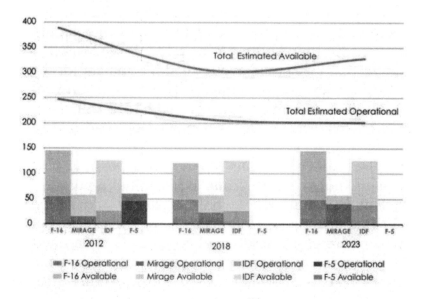

Source: *The Looming Taiwan Fighter Gap*, U.S.-Taiwan Business Council, October 1, 2012, p. 27, available from *www.us-taiwan.org/reports/2012_the_looming_taiwan_fighter_gap.pdf*.

Notes: Total estimated available signifies the total number of aircraft in the ROC fleet. Total estimated operational signifies total estimated available minus aircraft not expected to be available for use because of maintenance issues, aircraft out of service for planned modernization, and aircraft based in the United States for training.

Figure 9-2. Timeline of Estimated Total Fighters, 2012-23.

already outfitted with the missile; a mural at the 2011 Taipei Aerospace and Defense Technology Exhibition depicted the HF-3 sinking China's sole aircraft carrier. At the 2014 exhibition, the navy unveiled a prototype of a road-mobile launcher carrier for the HF-3.[26]

The HF-2E, on the other hand, is a surface-to-surface cruise missile designed to strike the Chinese mainland. It has a reported range of 600 kilometers. As recently as November 2012, former Taiwan deputy defense minister Andrew Yang told *Defense News* that "the U.S. is concerned about the development [of the HF-2E]. They are encouraging [China and Taiwan] to discuss the problem."[27] Even so, the HF-2E has entered active service and is deployed on road-mobile launchers.

There are conflicting reports on a possible new missile known as the Cloud Peak. Initial reports indicated that the missile would be supersonic, with a 1200-kilometer range allowing it to reach Shanghai and, perhaps, Chinese naval bases at Qingdao and Hainan island.[28] More recent reports described the Cloud Peak as a new land-based anti-ship cruise missile with a longer range than the HF-3.[29] Regardless of the precise nature of the Cloud Peak, Taiwan's missile program clearly remains a priority for the armed forces.

Taiwan's missile program flows in part from the MND's effort to develop "innovative and asymmetric" weapons and strategies to deal with the Chinese threat. As China knows, cruise missiles are a relatively low-cost capability against which it is costly and technologically difficult to defend. They are attractive to Taiwan's military for a number of reasons, including the fact that, in the event of a conflict, cruise missiles might be more likely than manned fighters to reach targets on the mainland. Strikes on critical Chinese command-and-control nodes could significantly impede PLA operations.

In addition, in fielding modern cruise missiles, Taipei conveys to Beijing that a war would not be confined to the island and surrounding waters. Cruise missiles allow Taipei to inflict costs on China, both by striking PLA targets and by bringing the war home for Chinese citizens. Deterrence, Taiwan believes, is enhanced as a result.

ISR.

To make effective use of all of these assets, Taiwan requires a suite of ISR systems. The pride of Taiwan's ISR is the military's new ultra-high-frequency (UHF) radar, which can track ballistic and cruise missiles and peer deep into China. A U.S. defense industry source told *Defense News* that "it's more of an intelligence collection system than a ballistic missile defense warning system" and that "Taiwan can see almost all of China's significant Air Force sorties and exercises from this radar." The radar is reportedly "capable of tracking 1,000 targets simultaneously."[30]

The new UHF radar, however, is just one piece of a larger picture. During the first decade of the 2000s, Taiwan made a concerted effort to develop its Po Sheng (command, control, communications, computers, intelligence, surveillance, and reconnaissance) system. According to Mark Stokes:

> the original Po Sheng concept . . . envisioned the installation of more than 750 data link terminals on most major weapons platforms that are integrated with joint [MND], army, air force, and naval operations centers.[31]

These links would provide for a common operating picture (COP), common operating environment, and enhanced command and control. Although resource limitations later restricted the scope of the system, one U.S. defense analyst has suggested that "Taiwan has the best common tactical picture in the world today, outside of the United States."[32]

Going forward, however, Stokes argues that Taiwan needs a "survivable network of sensors" for "pervasive and persistent surveillance," which might include earth observation satellites, "manned or unoccupied airborne sensors," and passive and active ground- and maritime-based sensors.[33] The MND shares a similar vision for its ISR capabilities. According to the most recent QDR, to enhance ISR, Taiwan will:

> effectively employ mid- and long-range electronic surveillance systems, extend ground, sea and air surveillance capabilities, integrate C2 systems, establish COP, and share battlefield information to enhance early warning capacity and battlefield transparency.[34]

AN UNCERTAIN FUTURE

Setting aside Taipei's difficulties in meeting its own defense spending targets, Taiwan faces two other significant obstacles to fielding the kind of force it envisions. First, the military's transition to an all-volunteer force faces major implementation problems and threatens to consume too many of the defense dollars that Taiwan does spend. Second, Taiwan continues to rely on defense articles from the United States at a time when Washington has a decreasing appetite for selling Taiwan the weapons it most needs and has an increasingly different vision than Taipei of Taiwan's optimal defensive strategy.

The All-Volunteer Force.

On its face, the rationale behind Taiwan's transition to an all-volunteer force makes sense. With Taiwan's low birthrate leading over time to a smaller labor force and considering the pull of the island's vibrant private sector on the smaller labor pool, a smaller active duty force has become increasingly attractive.

The MND has, moreover, reasoned that the development of high-tech weaponry allows for such a smaller force but does require a more highly trained one. Yet as the Legislative Yuan continued to reduce the amount of time that conscripts were required to serve in the military, maintaining such a highly trained force became increasingly difficult.

Taiwan is accordingly in the process of shrinking its military, which will come down from the current 215,000 personnel to between 170,000 and 190,000 by the end of 2019.[35] But the shift from a conscription system to a voluntary system is not going as well as hoped. For the first 11 months of 2013, recruitment levels were at only 30 percent of the target; "in infantry and armored units, the recruitment rates are even lower, at just 4 percent and 16 percent, respectively."[36]

The MND had planned on having an all-volunteer force in place by January 1, 2015, but that was postponed to 2017.[37] To boost recruitment, the ministry has plans to raise the starting monthly salary by more than 25 percent and to provide retirement benefits after 4 years of service rather than 10 years.[38]

In 2011, spending on personnel was at its highest since 2000. Moreover, after several years in which spending on personnel as a share of the total defense budget was well under 50 percent, that share was moving up again, hitting 45.36 percent in 2010 and 47.52 percent in 2011.[39] According to the U.S. Department of Defense (DoD), the rise in personnel costs is already "diverting funds from foreign and indigenous acquisition programs, as well as near-term training and readiness."[40]

Not only will Taiwan's transition to an all-volunteer force affect its ability to invest in new capabilities, but it also may negatively affect Taiwan's ability to defend against the most stressing of scenarios: an invasion by the PLA. The MND places an understandable emphasis on mobilizing reserves during such a scenario. The military's concept of "active duty force for strike and attack, reserve force for homeland defense" highlights the importance of the reserves, which is tasked with carrying out (alongside the active force) a key piece of Taiwan's defense strategy.

But maintaining an effective reserve force and promoting what the MND calls "all-out defense" is likely to become a greater challenge. Although all men of military age will continue to receive rudimentary training, the reserve force will be less well-trained than it was when all conscripts were required to serve on active duty.[41] While military personnel will serve in the reserves following their active duty service, these experienced servicemen and servicewomen will make up a smaller share of the reserve force as the active-duty force shrinks.

Not only does this look insufficient for building a force able to contend with a Chinese invasion, but it would also seem inadequate in the vein of MND's efforts to accumulate all-out defense capabilities. The QDR lists "all-out defense" as an important piece of the defense strategy's homeland defense mission:

> Continue to promote all-out defense education, cultivate the public's patriotism and support for national defense; coordinate interagency efforts to establish a robust all-out defense system; maintain capabilities of reserve force through mobilization and training to ensure rapid mobilization during peacetime and wartime.[42]

"All-out defense" includes an effort to improve civil-military relations, encourage an *esprit de corps* in the reserves, and promote willingness among the general population to support defense initiatives (and spending) and to actively contribute to defending the homeland in an emergency. Rather than achieving these goals, however, switching to an all-volunteer force and providing a bare minimum of reserve training may put these goals further out of reach.

U.S.-Taiwan Defense Relations.

From 1983 until the spring of 2001, when the United States held its last annual arms sales talks, American and Taiwan defense officials met once a year to discuss Taiwan's requirements and opportunities to acquire defense articles from the United States. The talks were halted during the first months of the George W. Bush administration, which had described China as a strategic competitor and wanted to put America's relationship with Taiwan on firmer footing. The end of the annual talks was envisioned as embodying an upgrade in U.S.-Taiwan relations.

The effect, however, was to remove the institutional impetus for regular arms sales to the island. Relieved of the requirement to discuss Taiwan's defense needs formally on an annual basis, the U.S. Government has frequently neglected to discuss them at all. The upshot is that since the last round of talks, new U.S. approvals of arms sales to Taiwan have been infrequent. In effect, the arms sales process has broken down over the last 13 years, and Taiwan has found it more difficult to purchase the defense articles it requires from the United States.

Taiwan's quest for new F-16 C/D fighter jets is illustrative. Taiwan first broached the subject with the George W. Bush administration in 2006. In what was then an unprecedented move, the administration refused to accept Taiwan's letter of request. Rather than approve or deny the request, the Bush administration would not even consider it. The Barack Obama administration adopted the same approach to Taiwan's F-16 C/D request, with the letter of request continuing to sit unopened in some Foggy Bottom inbox.

In 2011, U.S. President Obama did approve the sale of retrofits for Taiwan's existing F-16 A/B fighters, which the MND saw as a necessary complement to the acquisition of new fighters. But the decision to do so, while essentially continuing to ignore the question of new C/Ds for the island, demonstrated with surprising clarity a troubling development.

Under the Obama administration, the U.S.-China relationship has become, at the expense of Taiwan's defense needs, an increasingly central factor in decisions on U.S. arms sales to the island. The Obama administration's decision to upgrade Taiwan's F-16s while refusing to discuss new jets reflected a split-the-baby calculus: Washington would do the minimum for Taiwan, while keeping China happy. This sets a troubling precedent for future arms sales to the island.

Not only has the arms sales process largely broken down, but Taiwanese and American defense establishments are also disagreeing over the optimal strategy for Taipei to pursue and thus over what arms the island needs most. The provenance of this division appears to be, or is at least related to, U.S. Naval War College Professor William Murray's journal article recommending that Taiwan adopt a porcupine strategy. The article, which has been read widely at DoD, argues:

Rather than trying to destroy incoming ballistic missiles with costly PAC-3 SAMs, Taiwan should harden key facilities and build redundancies into critical infrastructure and processes so that it could absorb and survive a long-range precision bombardment. Rather than relying on its navy and air force (neither of which is likely to survive such an attack) to destroy an invasion force, Taiwan should concentrate on development of a professional standing army armed with mobile, short-range, defensive weapons. To withstand a prolonged blockade, Taiwan should stockpile critical supplies and build infrastructure that would allow it to attend to the needs of its citizens unassisted for an extended period. Finally, Taiwan should eschew destabilizing offensive capabilities, which could include, in their extreme form, tactical nuclear weapons employed in a countervalue manner, or less alarmingly, long-rang conventional weapons aimed against such iconic targets as the Three Gorges Dam.[43]

The strategy, which DoD has apparently endorsed in a less extreme form, calls for Taiwan to eschew expensive conventional capabilities—such as major surface and undersea combatants, fighter jets, and missile defenses—in favor of a ground-based, survivable, relatively inexpensive defensive force. The armed forces would focus on repelling an invasion and on homeland defense, to the exclusion of other missions such as counterblockade. Therefore, Taiwan would achieve deterrence through demonstrating to China that Taiwan would be a bitter pill to swallow, rather than through doing so in addition to holding at risk anything of value on the Chinese mainland.

While the argument for a porcupine strategy is not without merit, adopting it would severely limit Taiwan's options in the event of a crisis. It would

also require Taiwan to outsource the job of counter-
ing coercive uses of force short of invasion, when no
other country (including the United States) has made
a binding commitment to assume that responsibility.

Taiwan's own strategy, explicated in its QDR and
biennial national defense reports, is less narrowly focused. Taiwan's armed forces have identified a need
to counter China across the possible spectrum of coercive scenarios and in all domains of warfare. The
MND argues that it can adopt innovative asymmetric approaches to doing so, but it does not consider
hunkering down during a blockade or missile barrage
to be a realistic option, for reasons of deterrence or
domestic politics.

IS SECURITY WITHIN REACH?

Taiwan faces significant impediments to fielding a
force capable of carrying out its stated military strategy and, thus, to ensuring its security. The challenges
are few but significant: declining budgets, questions
about the viability of an all-volunteer force, uneven
and uncertain relations with the United States, and,
of course, the existence of an increasingly imposing
military force across the Taiwan Strait.

But no less an issue is the political-rhetorical problem in which Taiwan finds itself. To please Washington—Taiwan's only real security partner—and
to claim success in managing cross-Strait relations,
Taipei has had to argue that those relations are better
than they have ever been. Of course, in some respects
that is accurate. However, it has not lessened the actual military threat posed by mainland China to Taiwan.

Simply put, Taiwan's government must do a better
job of explaining that its policy of engaging with the

mainland does not eliminate the need to provide the island with an effective defense; indeed, only when Taiwan is secure can it, over the long run, engage China with confidence. Given China's assertive actions in the East and South China Seas and the more forceful and ambitious leadership of China's new president, Xi Jinping, addressing this shortcoming is more urgent than ever.

ENDNOTES - CHAPTER 9

1. This chapter was originally published as an essay on April 24, 2014.

2. Taiwan Relations Act, Public Law 96-8, 96th Cong., April 10, 1979.

3. Ministry of National Defense of the Republic of China, *National Defense Report 2013*, pp. 58, 60, 61, 66, available from *report. mnd.gov.tw/en/m/minister.html*.

4. *Ibid.*, p. 62.

5. Michael O'Hanlon, "Why China Cannot Conquer Taiwan," *International Security*, Vol. 25, No. 2, Fall 2000, p. 82.

6. Stockholm International Peace Research Institute (SIPRI) *Military Expenditure Database*, Stockholm, Sweden: SIPRI, available from *www.sipri.org/research/armaments/milex/milex_database*.

7. Shirley Kan, *Taiwan: Major U.S. Arms Sales Since 1990*, Washington, DC: Congressional Research Service, March 3, 2014, p. 34, available from *www.fas.org/sgp/crs/weapons/RL30957.pdf*.

8. SIPRI *Military Expenditure Database*.

9. Kan, pp. 33-34.

10. Anti-Secession Law, 10th National People's Congress, 3d Session, March 14, 2005, available from *www.china.org.cn/ english/2005lh/122724.htm.*

11. Executive Yuan, Republic of China Directorate-General of Budget, Accounting, and Statistics, *Statistical Yearbook of the Republic of China 2012*, Taipei, Taiwan, October 2013, available from *ebook.dgbas.gov.tw/public/Data/3117141132EDNZ45LR.pdf.*

12. Cindy Sui, "Changing Times Force Taiwan to Raise Welfare Spending," BBC News, April 24, 2013, available from *www. bbc.co.uk/news/business-22243977.*

13. Kan, pp. 57-58.

14. *National Defense Report 2013*, Beijing, China: Ministry of National Defense of the Republic of China, 2013, p. 68.

15. *Ibid.*, p. 83.

16. *Quadrennial Defense Review 2013*, Bejing, China: Ministry of Defense of the Republic of China, p. 36, available from *qdr.mnd. gov.tw/file/2013QDR-en.pdf.*

17. Chinese *National Defense Report 2013*, p. 274.

18. Chinese *Quadrennial Defense Review 2013*, p. 38.

19. *Ibid.*, p. 39.

20. *Ibid.*, p. 38.

21. Zachary Keck, "Taiwan Receives First 'Carrier Killer' Ship," *The Diplomat*, March 14, 2014, available from *thediplomat. com/2014/03/taiwan-receives-first-carrier-killer-ship/.*

22. Wendell Minnick, "Taiwan's Sub-Launched Harpoons Pose New Challenge to China's Invasion Plans," *Defense News*, January 6, 2014, available from *www.defensenews.com/article/20140106/DEFREG03 /301060013/Taiwan-s-Sub-launched-Harpoons-Pose-New-Challenge-China-s-Invasion-Plans.*

23. "Taiwan to Deploy 3 More PAC-3 Antimissile Batteries," *Want China Times*, April 26, 2013, available from *www.wantchinatimes.com/news-subclass-cnt.aspx?id=20130426000077&cid=1101*.

24. Aaron Mehta and Wendell Minnick, "F-16 Upgrade Dropped From US Budget Proposal, Sources Say," *Defense News*, January 27, 2014, available from *www.defensenews.com/apps/pbcs.dll/article?AID=2014301270023*; and Wendell Minnick and Aaron Mehta, "Taiwan Faces Tough Choices after US Cancels F-16 Upgrade," *Defense News*, March 8, 2014, available from *www.defensenews.com/article/20140308/DEFREG03/303080020/Taiwan-Faces-Tough-Choices-After-US-Cancels-F-16-Upgrade*.

25. Wendell Minnick and Aaron Mehta, "US Reassures Taiwan on Funding for F-16 Radar Upgrade," *Defense News*, March 23, 2013, available from *www.defensenews.com/article/20140323/DEFREG03/303230008/US-Reassures-Taiwan-Funding-F-16-Radar-Upgrade*.

26. Wendell Minnick, "Taiwan Displays New Missile Launch Vehicle," *Defense News*, August 14, 2013, available from *www.defensenews.com/article/20130814/DEFREG03/308140013/Taiwan-Displays-New-Missile-Launch-Vehicle*.

27. Wendell Minnick, "Q&A with Nien-Dzu Yang," *Defense News*, November 14, 2012, available from *www.defensenews.com/article/20121114/DEFREG03/311140011/Q-Nien-Dzu-Yang*.

28. Wendell Minnick, "Taiwan Working on New 'Cloud Peak' Missile," *Defense News*, January 18, 2014, available from *www.defensenews.com/article/20130118/DEFREG03/301180021/Taiwan-Working-New-8216-Cloud-Peak-8217-Missile*.

29. Wendell Minnick and Paul Kallender-Umezu, "Japan, Taiwan Upgrade Strike Capability," *Defense News*, May 7, 2013, available from *www.defensenews.com/article/20130507/DEFREG03/305060016/Japan-Taiwan-Upgrade-Strike-Capability*.

30. Wendell Minnick, "Taiwan's BMD Radar Gives Unique Data on China," *Defense News*, November 26, 2013, available from *www.defensenews.com/article/20131126/DEFREG03/311260013/Taiwan-s-BMD-Radar-Gives-Unique-Data-China*.

31. Mark A. Stokes, *Revolutionizing Taiwan's Security: Leveraging C4ISR for Traditional and Non-Traditional Challenges*, Arlington, VA: Project 2049 Institute, February 19, 2010, available from *www.project2049.net/documents/revolutionizing_taiwans_security_leveraging_c4isr_for_traditional_and_non_traditional_challenges.pdf*.

32. *Ibid.*

33. *Ibid.*

34. Chinese *Quadrennial Defense Review 2013*, p. 53.

35. "Size of Taiwan's Armed Forces to Be Cut Further," *Focus Taiwan*, January 20, 2014, available from *www.freesun.be/news/index.php/size-of-taiwans-armed-forces-to-be-cut-further/*.

36. Shang-su Wu, "Taiwan's All-Volunteer Military," *The Diplomat*, December 25, 2013, available from *thediplomat.com/2013/12/taiwans-all-volunteer-military/*.

37. Joseph Yeh, "MND Postpones Full Voluntary System to 2017," *China Post*, September 13, 2013, available from *www.chinapost.com.tw/taiwan/national/national-news/2013/09/13/388793/MND-postpones.htm*.

38. Cindy Sui, "Can Taiwan's Military Become a Voluntary Force?" BBC News, December 30, 2013, available from *www.bbc.co.uk/news/world-asia-25085323*.

39. Joachim Hofbauer, Priscilla Hermann, and Sneha Raghavan, *Asian Defense Spending, 2000–2011*, Washington, DC: Center for Strategic and International Studies, Defense-Industrial Initiatives Group, October 2012, available from *csis.org/files/publication/121005_Berteau_AsianDefenseSpending_Web.pdf*.

40. Office of the Secretary of Defense, *Annual Report to Congress: Military and Security Developments Involving the People's Republic of China 2013*, Washington, DC: U.S. Department of Defense, May 7, 2013, available from *www.defense.gov/pubs/2013_china_report_final.pdf*.

41. Once the transition is completed, there will continue to be universal training for young men of service age, but that training will last only 4 months. Beyond that, MND plans to provide 5 to 7 days of reserve force training twice yearly.

42. Chinese *Quadrennial Defense Review 2013*, p. 34.

43. William Murray, "Revisiting Taiwan's Defense Strategy," *Naval War College Review*, Vol. 61, No. 3, Summer 2008, available from *www.usnwc.edu/getattachment/Research---Gaming/China-Maritime-Studies-Institute/Published-Articles/Murray-NWCR-Summer08WEB1.pdf.*

CHAPTER 10

THE NORTH ATLANTIC TREATY ORGANIZATION's LAND FORCES: LOSING GROUND[1]

Guillaume Lasconjarias

The views expressed in this chapter are the author's own and do not necessarily reflect the opinions of the NATO Defense College or the North Atlantic Treaty Organization.

KEY POINTS

- Of all the service branches, allied land forces have borne the brunt of declining defense budgets.
- In recent years NATO's land forces have been professionalized and transformed, but their ability to carry out the various missions they might be tasked with is at risk because of a shortage of men and key materiel.
- Maintaining the operational experience and combat skills gained from deployments to Iraq and Afghanistan will be difficult, although the North Atlantic Treaty Organization (NATO) Response Force could, if properly structured, help ease that problem.

The state of NATO's land forces is something of a paradox. Although the alliance has no equal in terms of its gross domestic product, commands a wealth of human and social capital, and boasts the world's largest aggregate defense sector, NATO's land forces in particular have lost ground when it comes to their overall combat capacities.

In member states, the effects of the worldwide economic crisis on defense budgets have been compounded by dwindling public support for the continued commitment of national armed forces to apparently insoluble foreign conflicts. Nevertheless, as the alliance draws down its longest and costliest mission in Afghanistan, now is the time to review the lessons learned from a decade of sustained combat operations and to ensure they are implemented in time for the next major deployment. Overall, the idea is to shift from a "NATO deployed" to a "NATO ready" mode; the challenge, according to U.S. General Philip Breedlove, current supreme allied commander in Europe, is to maintain the operational excellence acquired over the past decade.[2]

At the strategic, operational, and tactical levels, land forces play a vital role, ensuring not only readiness for action at very short notice, but also ability to stay the course. Ground action is a major requirement in the three-step sequence of intervention, stabilization, and normalization and includes a wide range of missions from coercion to civil assistance.[3] To quote a former chief of staff of the French Army:

> Since war is mainly a question of controlling the population concerned . . . it will inevitably involve controlling the territories where these populations live — particularly urban areas, but also areas where ports and airports are situated. This means that troops on the ground will always be needed — and in sufficient numbers! — if one wishes . . . to obtain anything like a decisive victory. As a result, these troops on the ground will remain at the core of any future forces system.[4]

This is all the more problematic with the types of operations conducted in Iraq and Afghanistan, where civil war-like conditions raised the need for a very demanding level of territorial control only achievable by ensuring a high ratio of soldiers per inhabitant. Yet, with few exceptions, European armies continue a deflationary trend, moving more and more toward the status of "sample" or "bonsai" armies, which, in turn, risks breaking these smaller forces when deployed often and continuously.[5]

Some analysts see the U.S. pivot to the Asia-Pacific region as portending more generally the downsizing of both land force capacities and the role of land forces in future defense strategies. The growing emphasis on a new anti-access and area denial paradigm prioritizes air, naval, and amphibious operations. This idea is also gaining currency in Europe, where some politicians have even gone so far as to propose excluding land forces from any future operations.[6]

But many lessons learned over the past 2 decades of alliance operations lend support to the idea of maintaining credible land capabilities of an appropriate size and with a high level of technological sophistication. As Lieutenant General Frederick Hodges stated when NATO Allied Land Command Izmir (Turkey) became operational:

> Our tradition after every war has been repeating the mistake of reducing land forces to save money, believing that we can avoid casualties in future wars by relying more on air and sea power . . . and each time, we are required to hastily rebuild land forces to meet the threats the nation consistently fails to accurately anticipate.[7]

But whether NATO can avoid repeating this mistake in the face of eroding defense budgets and the uncertainty of a larger allied strategic vision is a question that remains both open and in need of answering sooner rather than later.[8]

TRANSFORMING LAND FORCES

The major challenge NATO member states face is to translate current security requirements into real operational capabilities.[9] Threats such as international terrorism and failed states have emerged alongside the more traditional threats posed by interstate tensions—a problem set that has reemerged with the Russian invasion of Ukraine.

To meet this complex set of security problems, since the end of the Cold War, NATO and its member states have made extraordinary efforts to transform their command and force structures, even in the face of declining budgets. Among the European NATO members, land forces have a number of common features:

- With few exceptions, conscription armies have been superseded by wholly professional forces.
- The "heavy" equipment for land forces of the Cold War period (for example, tanks and ground-based artillery) geared to an East-West conflict has given way to a new generation of high-tech equipment based on the principles of network-centric warfare.
- Command structures have been radically changed with the advent of standing multinational commands.
- New doctrines highlight the crucial role land forces play in stabilization operations and need to take a comprehensive approach involving political, military, and civilian assets when managing an armed intervention.[10]

But these changes have, in the face of fewer overall resources, come at a cost. No armed-service branch seems to have borne the brunt of these budget cuts more than the land forces, with troop numbers in some cases halved and equipment budgets slashed by two-thirds.[11]

The professionalization of NATO armies began in the 1990s. With the exceptions of Norway, Denmark, Estonia, Greece, and Turkey, most armies are now wholly professional. This meant a complete change not only in format, but also in the very way forces are structured. Thus, between 1996 and 2013, numbers in the French Army decreased from 268,572 to just more than 119,000. In the United Kingdom (UK), current troop numbers (99,800 in 2013) will be brought down to 82,000 by 2020.[12] For Germany's army, the trend is even clearer: the ongoing reform envisages a maximum of 61,000 soldiers. This compares with a total of 239,950 troops in 1996, of which 124,700 were conscripts.

The cuts are even more substantial among the former Warsaw Pact members. Joining NATO spurred them to place a priority on quality of land forces over quantity. For example, Polish land forces now total just 25 percent of the numbers they boasted just 20 years ago, while the Bulgarian army has shrunk from 50,400 (33,300 conscripts) in 1996 to 16,300 in 2013 and has eliminated all four tank brigades in favor of lighter infantry and more mobile mechanized units.[13] Throughout the alliance, units are being disbanded, facilities closed, and territorial defense structures reviewed.

This erosion of troop numbers limits the possible number of operational commitments a government can make and the size of the contingents that can be deployed. In France, for instance, the land forces envisaged under the 2014-19 military planning law will total 66,000 deployable soldiers, with a maximum commitment of 15,000 at any one time.[14] This means that, in little over a decade, France has gone from having a goal of being able to deploy 50,000 at any given time to a number less than two-thirds that figure.

The UK, meanwhile, aims to have a rapid-response force totaling five brigades, with the goal of having one brigade available at all times.[15] Based on a 36-month training and potential deployment cycle, this would allow the UK to undertake a brigade-level operation and two additional missions, one complex (up to 2,000 troops) and one simple (up to 1,000 troops).[16] Finally, in the case of the other major European power, Germany plans to have the ability to deploy up to 4,000 troops in two operational theaters and contribute about 1,000 troops for special operations, evacuation missions, or the NATO Response Force and European Union (EU) Battlegroups.[17] Given the potential manpower and resources available to both of these countries, the goals set on the number of deployable land forces are relatively modest.

For other nations, levels of potential operational commitment are less clearly formulated. In Italy, where land forces are still in the process of downsizing, projectable contingents are defined by the annual budget. Poland, after having been heavily involved in Iraq and Afghanistan, seems to be abandoning expeditionary capacity in favor of solely territorial deployment. In an official statement on August 15, 2013, President Bronisław Komorowski announced that the days of sending "Polish soldiers to the antipodes" were past.[18]

These limitations in deployable numbers are, theoretically, mitigated by the fact that national doctrines in most cases assess the legitimacy of any potential operation according to the number of democratic allies that would be involved. Thus, the general trend is in favor of coalition-based operations, either with a preferred partner (as in the case of the future Franco-British Combined Joint Expeditionary Force, born from the November 2010 Lancaster House Treaties) or by participating in an EU- or NATO-sanctioned operation.

Further efforts at mitigating the impact of shrinking force sizes include the introduction of unmanned robotic systems to replace personnel in certain functions — for example, the growing use of unmanned aerial vehicles (UAVs) and the deployment of robots for anti-improvised explosive device (IED) operations — and the greater reliance on smaller, albeit expensive, special forces to carry out specified missions.[19] In addition, reserve troops in some countries are growing in strength. Here, the British and Canadian models seem to be better established than in most other member states. However, the use of reserve troops is still subject to two major constraints: the actual availability of reservists and the quality of their training.[20]

But the fact remains that the nature of the conflicts experienced by NATO members since the end of the Cold War inescapably indicates that troops are still needed on the ground — and sometimes in considerable numbers if the overall mission is to be accomplished.

WHAT EQUIPMENT FOR WHICH OPERATIONS?

The economic crisis affects not only troop numbers but also the ability to introduce new equipment, with consequences for future force models. Further complicating the effort to upgrade platforms and weapon systems has been the constant pace of recent operations and those operations' particular equipment needs.

These considerations have led to radical choices in some nations. One example is European allies decommissioning their heavy tank units. While the modern tank is still a potent weapon and often overlooked as an efficient and relevant tool in current stability operations, Europe may only have 450 to 600 modern main battle tanks to be distributed among France, the UK, and Germany in a near future.

In the Netherlands, battle tanks have been totally eliminated. Interestingly, this decision was based not on an analysis of the operational environment but on budget-related considerations:

> The cuts imposed on the Royal Netherlands Forces . . . are an indirect consequence of the international economic crisis [and] there are no underlying strategic or political considerations other than the obvious need to recover economic health.[21]

Similarly, budget constraints have led Germany to halve its orders for Tiger attack helicopters (from 80 to 40) and those for the multirole NH90s (first from 122 to 80, then to 64). The budget forecast for Puma vehicles has been reduced from 410 to 342 while the number of Leopard 2 tanks will drop from 350 to 225.[22] The same trend can be seen in the UK, where 188 main battle tanks are to be cut from the land force, and the AS-90 self-propelling artillery system phased out.[23]

Squeezed by tight budgets, governments tend to view the modernization process as an opportunity to maintain national industries. France, Italy, the UK, and Germany, for example, each have their own model of heavy tanks. Meanwhile, in Turkey, where the first wave of modernization in the late-1980s included updating the oldest equipment with foreign buys, priority has now been given to the production and adoption of military equipment that has been domestically produced since the late-2000s. Pride of place goes to the $500 million Altay T battle tank project, with delivery scheduled to start in 2015.[24] Poland, too, has outlined a modernization effort it hopes to use to support and transform its defense industry.[25]

There is both an economic and political rationale for nations to keep specific manufacturing competences. No nation with a defense industry wants to give up such a resource, particularly when doing so could cause an increase in unemployment. Yet, for the period from 2010 to 2020, there are no fewer than 17 programs for the production of armored vehicles.[26]

Moreover, there is a tension between the desire to quickly update obsolescent materiel and the need to procure equipment over a period of years for budgetary reasons. This creates a stock of equipment from different generations. The British FV432 armored caterpillar-track troop carrier, first introduced in the 1960s, has received upgrades that would extend the vehicle's lifespan to 2020 and beyond and allow it to operate alongside much more modern equipment.[27] In addition, the need to test new materiel in real-life conditions further draws out the lifespan of their predecessors.

Land forces have in some cases also benefited from their combat deployment in Iraq and Afghanistan, obtaining equipment they would otherwise probably not have acquired as quickly. This can also result in what some experts call the hyper-specialization of land forces, with armies' procurement and logistic needs geared to the operational needs of a counter-insurgency (COIN) campaign.[28] Admittedly, modernization has also been positively informed by the practical experience of addressing urgent operational requirements, even at great expense. For example, when faced with new threats such as IEDs, allied land forces have chosen to move rapidly toward adopting better-protected vehicles.[29]

For many nations, the urgency to meet this need required purchasing vehicles from outside their own countries. Following its engagement in Afghanistan, the Netherlands purchased 76 Thales Australia Bushmasters between July 2006 and August 2009. Similarly, after seeing dozens of their armored vehicles destroyed by IEDs, the British contracted in November 2006 with Force Protection Inc., the producer of the American-made Cougar Mine-Resistant Ambush Protected Vehicle (MRAP) and supplier to the U.S. Marines Corps.

The Mastiff and Ridgeback variants of the Cougar MRAP are major items of expense at $623,000 and $600,000, respectively. One study estimates that the UK purchase of more than 750 vehicles under urgent operational procedures, mostly from U.S. suppliers, cost more than £260 million (€313 million).[30] Norway has approached BAE Systems for the upgrading of 103 CV90 combat vehicles, which were purchased from the mid-1990s onward, and the supply of 41 new CV90s between 2015 and 2017, accounting for a total outlay of about $1 billion.[31] In these cases, fleets could be too piecemeal or too small to be run effectively.

Vehicles purchased in response to particular needs can become the backbone of future fleets, compromising plans to modernize future land fleets with significantly different architectures.

Currently, European NATO member states are in a paradoxical situation, with certain items of capacity in excess (particularly infantry combat vehicles) and glaring shortfalls in other areas. These shortfalls are not tied to land forces alone, of course. For example, while difficulties related to strategic aerial transport should be corrected with continued procurement of the A400-M common European platform, many states depend either on extremely expensive outsourcing agreements with the private sector or on help from allies with C-5 and C-17 transport aircraft in their fleet inventory. Another well-known example is the shortage of UAVs, leading to almost exclusive dependence on the United States for battlefield intelligence from such platforms, as was the case in Libya and at the start of the French operation in Mali.

Helicopters are in particularly short supply. NATO's experience in Afghanistan has shown their importance in a broad range of missions including timely transport, convoy protection, fire support, and intelligence gathering. Operation UNIFIED PROTECTOR also underlined the full extent of their importance for strikes against Muammar Gaddafi's army in 2011 and in providing very close support for the rebel forces.[32]

Despite their proven utility, there are simply too few helicopters and crews in NATO armies. Note the British experience in Afghanistan: At the time of its initial deployment in Helmand in 2005-06, the UK command had eight Apaches and 10 utility helicopters. Parliament soon became concerned about the insufficient number and availability of helicopters for a range of different missions.[33] During the following years, the number of helicopters deployed grew continuously, reaching a peak of 35 in 2011. Even so, the Ministry of Defence had to outsource private helicopter services for delivery of supplies to their troops at Helmand bases, at an estimated cost of £4 million per month.[34]

Moreover, the helicopter fleets generally lack interoperability. European allies have two main types of new-generation utility helicopters, which are slowly being brought into service: the AgustaWestland Merlin HM1 and the NH90. A total of 58 Merlins have been ordered by three countries (Italy, Denmark, and Portugal), while 270 of the NH90s have been contracted for eight different countries.[35] Both models are more complex than their predecessors and capable of a wider range of tasks, resulting in higher costs and, in turn, smaller orders. At the same time, several former Warsaw Pact members (namely Hungary, the Czech Republic, Poland, and Slovakia) still use old Soviet Mil Mi-8s, Mi-17s, or Mi-24s. A declaration of intent was signed in 2009 to speed up work on compatibility between these helicopters and NATO's standard for its helicopters, but the economic crisis has hindered progress.[36]

Finally, there is no joint multinational helicopter command, and few NATO nations have the resources, including pilots, for complex air-land operations involving large numbers of helicopters. Moreover, the NATO standard of 180 hours of flying time per year is rarely met. Pilots in Italy, Germany, and Spain log an average of about 100 hours and Polish pilots fly only about 40 hours, which is complemented somewhat by training on simulators.[37] Here again, cost is a major consideration. According to one officer, the only way of ensuring adequate training for French pilots to deploy in operations was an appeal to superiors on the need to meet the NATO requirement.[38]

Recently, increased efforts have been made to enhance synergies, increase exchange programs for pilots, conduct joint exercises, and develop compatible doctrines. But gaps remain, including the pressing need for allied agreement on the requirements tied to the future development of a heavy-transport helicopter.

LESSONS FROM THE PAST DECADE

After a decade of crisis management and peacekeeping, the return to war in the early-2000s has had a lasting effect on allied land forces. Whereas the contingents deployed to the Balkans—the Implementation, Stabilization, and Kosovo Forces—thought in terms of stabilization and reconstruction, the turning point came with the engagement in Iraq, and then again in Afghanistan. In both cases, existing military resources and doctrines were ill-suited to the complexities of those environments and no longer attuned to the demands of the kind of asymmetric warfare allied land forces faced.

Forces were maladapted to the specificities of this new warfare on two counts. First, there was an urgent need to update materiel with a view to protecting forces. Second, there was a pressing need to face the challenge of a different environment. From a doctrinal perspective, this required an understanding of how forces were to be used in COIN operations, which was unfamiliar territory to all but a few allies. This prompted urgent work on new field manuals by the allies, borrowing heavily from the U.S. Army's 2006

Field Manual 3-24 Counterinsurgency.[39] NATO got into the act as well, belatedly publishing the *Allied Joint Publication (AJP) 3.4.4, Allied Joint Doctrine for Counterinsurgency (COIN)* in February 2011. While describing the need for a "comprehensive approach" involving multiple civilian, political, and military organizations and agencies in the COIN effort, the document focuses, not surprisingly, on the political role played by land forces in this asymmetric environment.[40]

Undoubtedly, the most significant takeaway from allied armies' experience in Iraq and Afghanistan is that the expertise of land forces has to be extended to new domains. Originally devised simply to coerce the enemy, armies in these environments will now have responsibilities across a broad spectrum and, hence, will need additional capabilities to address them effectively. This means combining multiple approaches to accomplishing strategic goals, emphasizing decentralized command and control, effectively training indigenous forces and, in general, being willing to work in a joint forces setting and in conjunction with other ministries to provide security in distinct regions, reassure local populations, and rebuild social and governing institutions to reattach the population to a legitimate political authority.

Another key capability that has already proved its worth is the security forces assistance mission. Successful training missions can help prevent crises, help failed states recover, shorten intervention times, and facilitate the withdrawal of foreign allied forces.

From 2004 to 2011, NATO Training Mission-Iraq trained more than 15,000 personnel with less than 200 trainers. The NATO Training Mission-Afghanistan (NTM-A) has been even more successful. At its peak, NTM-A employed 2,800 trainers and was working with 34,000 Afghans across 70 training sites. Even today, there are 1,900 personnel from 39 nations, and on any given day, more than 20,000 personnel are being trained.

Before NTM-A was operational, only a third of Afghan soldiers met NATO marksmanship standards. Today, that figure is 97 percent. The NTM-A's desired end state is one of Afghan ownership. Today, 95 percent of all conventional operations and 98 percent of all special operations are conducted by Afghan military and security personnel. While allied and partner forces have helped create a space in which a fledging army in Afghanistan could get its feet on the ground, it is the training mission that will ultimately provide the Afghans with the capacity to secure their own nation. All of this requires boots on the ground.

THE NATO RESPONSE FORCE

After years of discussion regarding the rationale, effectiveness, and role of the NATO Response Force (NRF), it has the potential to be a catalyst for maintaining a modern, allied land force.[41] Conceived as a multinational joint force, the NRF is intended to provide a robust and rapidly deployable coalition force to meet a range of missions, from the evacuation of civilians to a high-intensity engagement. The makeup of the NRF is straightforward: individual nations make contributions to the force structure for 1 year while a multinational rapid response command is kept on standby. Once a command's readiness is certified, it is set to perform the tasks entrusted to it by the alliance. The NRF is also important as a setting for major live exercises. The most recent, Steadfast Jazz 2013, brought together some 6,000 troops in Poland. While there is still considerable discussion about the NRF's potential uses, the force is nevertheless a formidable resource that ensures basic levels of training, available manpower, and military capacity.

However, the present state of the NRF reflects the strengths and weaknesses of the current state of allied armed forces. For 2014, the land component of the NRF involves 12 allied countries and one partner state, Ukraine. The challenge is keeping the dedicated forces operationally prepared for possible deployment at short notice. Here, the main problem is that the NRF command is not directly in charge of the units that can be assigned to it, which are spread out among contributing nations. In addition, there is the problem that certain capacities—such as helicopters or UAVs—are missing or insufficient in number.

According to some analysts, however, the real issue is somewhat different. The NRF is viewed by some in the alliance as not so much a resource for actual use as a test bed for increasing interoperability among alliance partners. From this viewpoint, the NRF is about allied forces getting acquainted with each other, training under shared procedures, and exploring new operational possibilities. The NRF thus offers a platform for operational convergence, promoting a common spirit through a network of certified units.[42]

The NRF is particularly important in relation to the upcoming withdrawal from Afghanistan and the apparent end of alliance forces' major foreign engagements. It can help ensure that standards are maintained and that some percent of the forces are kept in a state of readiness. Participation in exercises and the mandatory certification process also ensure that financial resources are earmarked by governments. Finally, the NRF process allows alliance forces the unique opportunity to experiment with new technologies and operational concepts, with an important trickle-down effect on their respective militaries.

CONCLUSION

The ongoing reforms in European armies were initially nurtured in the 1990s by a number of illusions, starting with the idea that land force-intensive wars were a thing of the past and, in turn, that peacekeeping or peacebuilding missions were going to predominate future land force use. These reforms thus faced the challenges of governments giving even greater priority to domestic programs and of a major economic crisis in recent years. The result: cost-saving became the principal consideration, with the brunt of the related cuts mostly borne by member states' land forces. The low level of overall strength they have now reached leaves them weakened and even jeopardizes their overall coherence. Facing on the one hand demanding deployments over the past decade and, on the other, continual attempts at reorganization and transformation, it is hardly surprising that, while Europe's allies have an abundance of manpower, in 2012 they had some 1.56 million soldiers under arms, but less than 5 percent really deployable.

Concentrating on core combat capabilities at the expense of combat support functions such as logistics and engineering initially made it possible for defense ministries to continue making substantial contributions to alliance military missions. However, the shortfalls are now apparent, with the major European land powers (France, the UK, and Germany, followed by Italy and Spain) struggling to maintain the necessary component parts that constitute a land force capable of joint air-land operations. They are now mere "sample" forces, kept at a level of numbers and materiel that makes them increasingly irrelevant as individual nation-state combatants.

How might the future look, then? One possible model now being touted is that of the 2013 French-led operation in Mali. With Paris in the lead and its forces with resources in locations in and around the zone of conflict, NATO and EU allies were able to support the French operation by providing additional needed capabilities such as UAVs, airlift, and intelligence. But this model might not be so easily duplicated, requiring a political leadership, like the French, willing to assert its strategic will and have sufficient deployable combat power both to address the contingency and be the core element around which allies can help fill in missing operation pieces. Whether this coalition of the willing can be a true model for future force planning is far from clear.[43]

But as Western armies have discovered in Iraq, Afghanistan, and Lebanon with Hezbollah, dealing with irregular forces is operationally difficult, complex, and resource intensive. This has generated reluctance among both politicians and publics to deploy their forces, especially their land forces, far afield and for a long time. Not surprisingly, this has, in turn, led governments and strategists to look to high-tech weaponry and special forces conducting quick in-and-out strikes to carry the load. But, like the Mali model, there are limits to what U.S. General H. R. McMaster calls "global swat teams" can accomplish.[44]

Nevertheless, a fundamental reality of war and politics remains: at times, only the physical presence of land forces can offer the hope of resolving a crisis and stabilizing the situation on the ground. These forces' adaptability and capacity to engage in a broad range of missions is thus part of the resources that government leaders must still have if they hope to meet their respective countries' larger foreign policy goals.[45] War weary or not, NATO members are ignoring tactical and strategic realities—and, indeed, history—if they believe that continuing to drain their land forces of numbers and capabilities is either wise or sustainable.

ENDNOTES - CHAPTER 10

1. This chapter was originally published as an essay on June 4, 2014.

2. Jonathan Marcus, "NATO Commander Philip Breedlove on Post-Afghan Future," BBC News, July 3, 2013, available from *www.bbc.co.uk/news/world-asia-23157256*.

3. Centre de Doctrine d'Emploi des Forces, "Stratégie, 'opératique' et tactique: La place des forces terrestres" ("'Operatic' Strategy and Tactics: The Place of Ground Forces"), December 2005, available from *www.cdef.terre.defense.gouv.fr/publications/anciennes-publications/articles-sur-le-retex/retex-doctrine/doctrine-07*; Charles Krulak, "The Strategic Corporal: Leadership in the Three Block War," *Marines Magazine*, January 1999, available from *www.au.af. mil/au/awc/awcgate/usmc/strategic_corporal.htm*; and Rupert Smith, *The Utility of Force: The Art of War in the Modern World*, London, UK: Allen Lane, 2005.

4. Elrick Irastorza, "Forces armées: peut-on encore réduire un format 'juste insuffisant'?" ("Armed Forces: Can a "Just Enough" Format Be Further Reduced?"), Paris, France: French Senate, Commission on Foreign Affairs and Defense, July 18, 2012, available from *www.senat.fr/rap/r11-680/r11-680_mono.html*.

5. Christian Mölling, "Europe without Defense," *Stiftung Wissenschaft und Politik*, November 2011, available from *www.isn.ethz. ch/isn/Digital-Library/Publications/Detail/?id=135191&lng=en*.

6. Ben Barry, "The Age of Gloom? Implications for Key NATO Armies," *Land Warfare*, Vol. 33, No. 4, July 2013, p. 3, available from *www.rusi.org/downloads/assets/201307_NB_Barry.pdf.*

7. Frederick Hodges, "Ensuring That Land Forces Remain Decisive for NATO," *Army Magazine*, May 2013, pp. 51-54, available from *www.ausa.org/publications/armymagazine/archive/2013/05/ Documents/Hodges_May2013.pdf.*

8. The economic crisis has accentuated the already-marked reductions in defense budgets. Currently, all European NATO members except Poland are following this trend. Average defense expenditure for member states is 1.6 percent, with considerable disparities. According to the latest official NATO figures, 2 countries (Greece and the UK) meet the objective of spending 2 percent of gross domestic product on defense (excluding pensions); eight countries (Albania, Croatia, Estonia, France, Norway, Poland, Portugal, and Turkey) spend 1.5 to 2 percent, 12 countries spend 1 to 1.5 percent (Belgium, Bulgaria, Germany, Czech Republic, Denmark, Hungary, Latvia, the Netherlands, Romania, Slovakia, and Slovenia), and only three countries spend less than 1 percent on defense (Lithuania, Luxembourg, and Spain). See North Atlantic Treaty Organization, Public Diplomacy Division, "Financial and Economic Data Relating to NATO Defence," press release, April 13, 2012, available from *www.nato.int/cps/en/natolive/news_ 85966.htm.*

9. Olivier de France and Nick Whitney, *Étude comparative des livres blancs des 27 états membres de l'Union européenne: pour la définition d'un cadre européen* (*Comparative Study of White Papers of the 27 Member States of the EU: To Define a European Framework*), Paris, France: Institut de Recherche Stratégique de l'Ecole Militaire, 2012, available from *www.defense.gouv.fr/content/download/185008/2037037/file/Etude%2018-2012.pdf.*

10. This partly reflects information given by Antonio Missiroli, ed., *Enabling the Future: European Military Capabilities 2013-2025: Challenges and Avenues*, Report No. 16, Pretoria, South Africa: Institute for Security Studies, May 2013, p. 9, available from *www. iss.europa.eu/uploads/media/Report_16.pdf.*

11. This is particularly true in Southeast Europe, according to Anton Bebler, *Sodobni vojaški izzivi* (*Security Challenges in South Eastern Europe*), Ljubljana, Republic of Slovenia: Ministry of Defence, November 2013, pp. 39-50, available from *www.slovenskavojska.si/fileadmin/slovenska_vojska/pdf/vojaski_izzivi/svi_15_3.pdf*.

12. The British model nevertheless prioritizes the role of 30,000 trained reservists within an integrated army, bringing total numbers to 112,000. See UK Ministry of Defence, *Transforming the British Army: An Update*, July 2013, p. 18, available from *www.army.mod.uk/documents/general/Army2020_Report_v2.pdf#search=July2013*.

13. "Chapter 4: Europe," *The Military Balance 2012*, London, UK: International Institute for Strategic Studies IISS), March 2012, pp. 112, 143.

14. The military planning law also entails a reduction of land forces by 6,000 and the loss of a joint-force brigade. See Gwenegan Bui, "Avis relatif à la programmation militaire pour les années 2014 à 2019 et portant diverses dispositions concernant la défense et la sécurité nationale" ("Notice on the Military Planning for the Years 2014-19 and Containing Various Provisions on Defense and National Security"), No. 1540, French National Assembly, November 2013, pp. 18-19, available from *www.assemblee-nationale.fr/14/rapports/r1540.asp*.

15. *Transforming the British Army*, p. 6.

16. See Her Majesty's Government, *Securing Britain in an Age of Security: Strategic Defence and Security Review*, October 2010, pp. 19, 24, available from *www.direct.gov.uk/prod_consum_dg/groups/dg_digitalassets/@dg/@en/documents/digitalasset/dg_191634.pdf*.

17. Federal Ministry of Defence, *Die Neuausrichtung der Bundeswehr* (*The Reorientation of the Bundeswehr*), March 2013, p. 39, available from *www.bmvg.de/resource/resource/MzEzNTM-4MmUzMzMyMmUzMTM1MzMyZTM2MzIzMDMwMzAzMD-MwMzAzMDY4NjY2NjM2NmM2Mjc5NjQyMDIwMjAyMDIw/Die%20Neuausrichtung%20der%20Bundeswehr_M%C3%A4rz%20 2013_final_Internet.pdf*.

18. Patrycja Bukalska, "The Polish Army: New Armor for a New Mission," Visegard, Hungary, October 5, 2013, available from *www.visegradrevue.eu/?p=1840*.

19. This trend was reflected in the establishment of the NATO Special Operations Headquarters in June 2007. According to its website, its "co-location with the Allied Command Operations (ACO) strongly affirms the NSHQ's function and support in providing Special Operations advice to SACEUR." See NATO, "NSHQ: About," available from *www.nshq.nato.int/nshq/about/*.

20. Hence the need for innovative reforms in this area. See Guillaume Lasconjarias, *Send the Reserve! New Ways to Support NATO through Reserve Forces*, Research Paper No. 99, Rome, Italy: NATO Defense College, 2013, available from *www.ndc.nato.int/download/downloads_fr.php?icode=398*.

21. Interview with Peter van Uhm, *Défense et Sécurité Internationale*, *(International Defense and Security)*, No. 74, October 2011, pp. 40-43.

22. Federal Ministry of Defence, Die Neuausrichtung der Bundeswehr, pp. 35-45.

23. Andrew Dorman *et al.*, *The Implications of Military Spending Cuts for NATO's Largest Members*, Washington, DC: The Brookings Institution, July 2012, p. 12, available from *www.brookings.edu/research/papers/2012/07/military-spending-nato-odonnell*.

24. With initial delivery of 250 tanks and a subsequent batch of an additional 750, the aim is to replace the venerable M48 Patton.

25. Dominik P. Jankowski, *Beyond Air and Missile Defense: Modernization of the Polish Armed Forces*, Issue Brief No. 132, Washington, DC: Center for European Policy Analysis, September 5, 2013, available from *cepa.org/content/beyond-air-and-missile-defense-modernization-polish-armed-forces*. See also Andrew A. Michta, "Polish Hard Power: Investing in the Military as Europe Cuts Back," *AEI National Security Outlook*, December 2013, available from *www.aei.org/outlook/foreign-and-defense-policy/defense/nato/polish-hard-power-investing-in-the-military-as-europe-cuts-back/*.

26. Acquiring equipment from national manufacturers rather than off the shelf or through participation in multinational programs is not wholly disadvantageous, however. Consider the French Army's armored combat infantry vehicle (véhicule blindé de combat d'infanterie), which, at a unit cost of €4 million, is six times more expensive than its predecessor. However, the new vehicle proved its worth during Operation SERVAL in Mali, enhancing troops' effectiveness by offering increased protection and fire support. See Jacques Marie Bourget, "'On y allait pour leur casser la figure', déclare le Général Barrera" ("'We Went There to Hit Them Hard,' Said General Barrera"), *Mondafrique*, November 22, 2013, available from *www.mondafrique.com/lire/decryptages/2013/11/22/on-y-allait-pour-leur-casser-la-figure-declare-le-general-barrera*.

27. Jean-Jacques Mercier, "La reconstruction des APC, deux approches" ("The Reconstruction of APC, Two Approaches"), *Défense et Sécurité Internationale*, No. 67, February 2011, pp. 93-97. The British Army still has more than 1,500 FV432s, including almost 1,000 that were involved in this upgrade between 2005 and 2011 at a cost of nearly £230 million.

28. See Mark Philips, *Exercise Agile Warrior and the Future Development of UK Land Forces*, RUSI Journal, May 2011, p. 15, available from *www.rusi.org/downloads/assets/agilewarrior.pdf*; and Barry.

29. The procurement procedures are very cumbersome in some nations. In France, according to official statistics issued by the integrated structure in charge of ensuring operational readiness of materiel for land operations (the SIMMT), the lead time is "7 months, on average, to purchase a complete item of materiel by adapted procedure . . . and 18 months on average for an agreement regarding maintenance." See "Le bilan contractuel" ("The Contract Review"), *Les échos de la SIMMT*, (*The Echoes of the Integrated Structure of the Operational Maintenance of Army Equipment [SIMMT]*), No. 6, February 2012, pp. 2-5, available from *www.asso-minerve.fr/wp-content/uploads/2011/02/Les_Echos_de_la_SIMMT_n-_6.pdf*.

30. Chris Maughan, "The Impact of UORs on the UK Defence Industry," *RUSI Journal*, February 2009, pp. 90-93, available from *www.rusi.org/downloads/assets/RDS_Maughan_Feb09.pdf*.

31. Stephen Gilbert, "Improving the Survivability of NATO Ground Forces," North Atlantic Treaty Organization Parliamentary Assembly, October 13, 2013, p. 5, available from *www.nato-pa. int/shortcut.asp?FILE=3302*.

32. Apache attack helicopters operating from the HMS *Ocean* logged about 50 sorties, firing 99 Hellfire missiles and 4,800 30-millimeter shells against a total of 116 targets. French helicopters launched from landing platform docks destroyed more than 550 targets in 40 sorties. The Tiger, operating mostly at night, fired 1,500 rockets, and the Gazelle shot 430 high subsonic optical remote-guided, tube-launched missiles. See *Lessons Offered from the Libya Air Campaign*, London, UK: Royal Aeronautical Society, July 2012, pp. 10-11, available from *www.aerosociety.com/Assets/Docs/ Publications/SpecialistPapers/LibyaSpecialistPaperFinal.pdf*.

33. Parliament, House of Commons Defence Committee, "The UK Deployment to Afghanistan: Government Response to the Committee's Fifth Report," Sess. 2005-06, June 13, 2006, p. 5, available from *www.publications.parliament.uk/pa/cm200506/ cmselect/cmdfence/1211/1211.pdf*.

34. Caroline Wyatt, "Crucial Role of Helicopters in Afghanistan," BBC News, March 2, 2010, available from *news.bbc.co.uk/2/hi/ uk_news/8545695.stm*.

35. It should be noted that delivery times are long, and cancellations or reductions could further limit the final total. For France, delivery of NH90s will continue until 2024, meaning that almost-obsolescent helicopters will have to be kept until 2030. See French National Assembly Committee on National Defense and Armed Forces, Compte rendu n°13 (Report No. 13), October 16, 2013, available from *www.assemblee-nationale.fr/14/cr-cdef/13-14/ c1314013.asp*.

36. *Enhancing NATO's Operational Helicopter Capabilities: The Need for International Standardisation*, Kalkar, Germany, Joint Air Power Competence Center, August 2012, p. 10, available from *www.japcc.org/wp-content/uploads/Helicopter_Capabilities_web.pdf*.

37. *Ibid.*, p. 7.

38. Private interview with a French officer in Marseille, France, April 15, 2013.

39. *Field Manual (FM) 3-24: Counterinsurgency*, Washington, DC: Headquarters, Department of the Army, June 16, 2006, available from *www.fas.org/irp/doddir/army/fm3-24fd.pdf*.

40. See AJP-3.4.4, pp. 1-5, available from *info.publicintelligence.net/NATO-Counterinsurgency.pdf*:

It should invigorate existing processes and strengthen relationships at the joint, inter-agency and multinational levels. A comprehensive approach should also consider actors beyond government, such as NGOs, IOs and others all conduct activities that have a bearing on the overall outcome. This is particularly relevant for land forces at all levels where they should expect to operate alongside these actors.

41. For more on the NRF and related topics, see Guillaume Lasconjarias, *The NRF: From a Key Driver of Transformation to a Laboratory of the Connected Forces Initiative*, Research Paper No. 88, Rome, Italy: NATO Defense College, January 2013, available from *www.ndc.nato.int/download/downloads.php?icode=363*.

42. Anthony King, *The Transformation of Europe's Armed Forces: From the Rhine to Afghanistan*, New York: Cambridge University Press, 2011, p. 92.

43. Sidney Freedberg, "The Army Cuts: 3 Truths, 4 Fallacies," *Breaking Defense*, February 24, 2014, available from *www.breaking-defense.com/2014/02/the-army-force-cuts-3-truths-4-fallacies*.

44. *Ibid*.

45. The tensions and violence in the Central African Republic are a case in point, requiring France to deploy several hundred extra troops.

CHAPTER 11

UNITED KINGDOM HARD POWER: STRATEGIC AMBIVALENCE[1]

Paul Cornish

KEY POINTS

- On a comparative scale, the United Kingdom (UK) remains a significant military power with significant operational experience.
- Since the Cold War's end, Britain has attempted to adapt to a more complex set of security problems while simultaneously cutting force structure.
- As a result, the UK strategic field of vision has shrunk: it has increasingly adopted a preference for military operations that are far away, fairly small, or relatively brief.

Over the past 2 decades, the relatively settled animosity of the Cold War has been replaced by a range of diverse, complex, and often very urgent security threats and challenges, albeit of a lesser scale. The 21st century is not proving to be as dangerous as some had feared, but neither is it as stable as many would wish.

As well as uncertainty, diversity, complexity, and urgency, there are scarcity and austerity: national strategy must compete for scarce resources and must take its share of continuing retrenchment in public spending. Pity the strategist expected to make durable and coherent decisions under such circumstances.

Yet national strategy is not a fair-weather activity; decisions must be made and cannot be postponed until more favorable circumstances arise. Among the most complicated of these decisions are those that concern a country's military force structure—the material basis of its so-called hard power. These decisions require a reasonably settled "threat picture" around which to construct a military architecture; sufficient flexibility to deal with unanticipated threats and challenges; political, public, and media support if the decisions are to be maintained over time; advanced technological knowledge; and, finally, a very high level of political and institutional confidence in spending vast amounts of public money on platforms and equipment that might well be in service for decades.

The purpose of this chapter is to gauge the strategic quality and vitality of UK hard power. I argue that it is not currently in the best of health and for reasons that are often misunderstood. There is widespread concern that UK armed forces have recently been reduced too far and, furthermore, that these reductions are symptomatic of a deep malaise in the British national psyche: a form of strategic "declinism," perhaps.

I do not believe that a narrow assessment of the size, shape, and capability of a country's armed forces reveals all that is to be said about its strategic ambition and, indeed, its hard power. UK armed forces are certainly smaller in 2014 than they were in 1945 (at the end of World War II), in 1982 (at the beginning of the Falklands War), and in 1989 (at the beginning of the end of the Cold War), but these comparisons tell us little. As well as assessing size and capability, a complete analysis of a nation's hard power requires an answer to one further question: what will it be for?

Since the end of the Cold War, UK military power has become less concerned with the defense of the country's territory (including its overseas possessions), its airspace, and its territorial waters and is much more concerned with addressing strategic challenges to the UK at those challenges' point of origin. In his introduction to the *1998 Strategic Defence Review* (SDR), then-secretary of state for defense George Robertson argued that "[i]n the post-Cold War world, we must be prepared to go to the crisis, rather than have the crisis come to us."[2] This notion of self defense at arm's length subsequently became a leitmotif in UK national strategy, attracting bipartisan consensus.

The first and most obvious indicator of the strength and scope of UK hard power is the amount Her Majesty's Government spends on the country's military force posture. The next indicator is capabilities: the platforms, equipment, weaponry, and personnel necessary for military commitments and for what are now described as expeditionary operations. Hard power is also reputational, concerned with the nation's experience with military operations and those operations' effectiveness. Finally, hard power is the expression of foreign policy outlook and strategic intent.

Although national strategy is concerned with the future, a nation's strategic posture—particularly its hard power—cannot develop in an instant and must evolve over time. This chapter covers the 15-year period from July 1998 to December 2013, beginning with the publication of the newly elected Labour Party government's SDR, which marked the beginning of a new, genuinely post-Cold War era in strategic UK thinking.

UK MILITARY EXPENDITURE

Military expenditure can be surprisingly difficult to track, as accounting procedures change from time to time. Nevertheless, it offers some indication of a country's intentions and seriousness.

The data in Figure 11-1 do not paint a picture of radical decline in UK military expenditure since 1998, nor even gradual decline, for that matter. On the contrary, annual spending has been held at a healthy level, allowing the UK to maintain its position as one of the world's top military spenders, even in the straitened economic circumstances following the 2008 financial crisis. Military expenditure as a percentage of gross domestic product (GDP) has likewise remained above the North Atlantic Treaty Organization (NATO) benchmark: together with the United States and Greece, the UK is one of only three NATO allies to have held to the 2006 commitment to spend a minimum of 2 percent of GDP on defense.

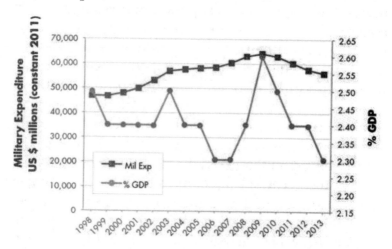

Source: Stockholm International Peace Research Institute (SIPRI), *Military Expenditure Database*, available from *www.sipri.org/ reesearch/armaments/milex/milex_database.*

Notes: Beginning in the late-1990s, UK defense budgeting was moved over a period of years from cash accounting to resource accounting and budgeting. The percentage of GDP figure includes military pensions, in accordance with NATO's 2004 revised definition of military expenditure.

Figure 11-1. Military Expenditure, 1998-2013.

This is not to suggest that discussion of UK military expenditure has been entirely free from contention. Since the transfer of power from the Labour to Coalition government in 2010, the UK defense debate has been dominated by the discovery of a so-called black hole in the defense budget: an unfunded liability of committed expenditure (largely on new equipment) between 2010 and 2020. A figure of £38 billion (roughly equivalent to the UK annual defense budget) is most often cited, although there are uncertainties as to how that sum was calculated.

Nevertheless, in September 2011, Defense Secretary Liam Fox announced that the shortfall had almost been eliminated, and just months later, Philip Hammond, Fox's successor, was reportedly confident that the black hole had been entirely eliminated and that the government would indeed be able to fund the Future Force 2020 (FF2020) modernization program that was announced in the 2010 *Strategic Defence and Security Review* (SDSR).[3]

In January 2013, the Ministry of Defence (MoD) announced that it was now in a position to commit as much as £160 billion over 10 years to a defense equipment plan "that will enable the MoD to deliver Future Force 2020." In Hammond's words:

> This £160 billion equipment plan will ensure the UK's Armed Forces remain among the most capable and best equipped in the world, providing the military with the confidence that the equipment they need is fully funded.[4]

Hammond's confidence is open to question, however. In the first place, some defense industrialists and acquisition analysts are concerned that the MoD has simply replaced the irresponsibility of overspending with the neurosis of underspending. By one account, successive "reform," "transformation," and "efficiency" programs have eroded the skills, morale, and strength of the MoD's civilian staff, resulting in "a department of state which, rather unusually, is both short of money and, with reduced personnel, lacking the capacity to spend the budget allocated to it."[5]

Furthermore, certain budgetary assumptions upon which the FF2020 construct was based were challenged by a series of cuts and adjustments made in the 2013 UK government spending review. Not the least of these was the decision to depart from past practice in the funding of operational military deployments. Whereas in the past such costs had come from the treasury's contingency reserve, henceforth, the MoD's main budget will be liable for as much as 50 percent of operational costs.

With a recent assessment suggesting that the overall cost of UK involvement in operations in Afghanistan and Iraq could be close to £30 billion, operational costs might represent a very significant new charge on the defense budget.[6] As Andrew Dorman and I have argued:

> for these financial reasons alone it is difficult to see how the structure and goal of FF2020, as published in SDSR 2010, can be considered affordable and therefore achievable-unless, as some world-weary commentators suggest, 2025 is to become 'the new 2020'.[7]

UK MILITARY CAPABILITIES

The picture is less encouraging when military capabilities are considered. Figures 11-2, 11-3, and 11-4 show trends in land, naval, and air capabilities, respectively. Each figure shows the regular (full-time professional) personnel strength of its featured force and the key hard-power expeditionary capabilities in each case.[8]

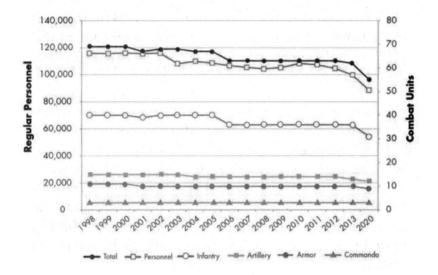

Source: "UK Defence Statistics Compendium," London, UK: UK Ministry of Defence, 1998-2013, available from *https://www.gov.uk/government/statistics/uk-defence-statistics-compendium-2013.*

Figure 11-2. UK Land Forces: Strength and Deployable Units, 1998-2013.

The principal land force capability is the battalion-sized unit that could form the basis of a deployable battlegroup: the army's armored regiments and infantry battalions, together with Royal Marine commandos. Artillery tactical fire support would also be essential to any operational deployment.

Figure 11-2 shows a reduction of land-force personnel by approximately 14 percent between 1998 and 2013, while the number of deployable battalion-sized combat units (including Royal Marine commandos) decreased by some 10 percent.[9] Although these reductions could scarcely be described as radical, they are certainly significant, particularly when lengthy operational commitments are undertaken: the fewer the units, the greater the frequency of deployment, with attendant effects on morale, family life, and retention. Projections to FF2020 show a further 11 percent reduction in both personnel and deployable units from 2013.

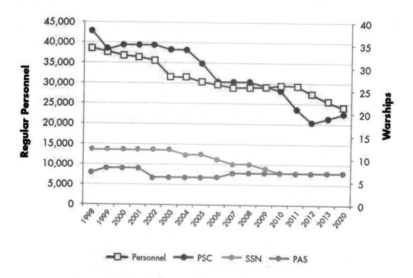

Source: "UK Defence Statistics Compendium," London, UK: UK Ministry of Defence, 1998-2013, available from *https://www.gov.uk/government/statistics/uk-defence-statistics-compendium-2013*.

Figure 11-3. Deployable Naval Forces, 1998-2013.

As well as the strength of regular naval personnel, Figure 11-3 shows the number of warships in three categories: attack submarines (SSN); aircraft carriers, destroyers, and frigates (principal surface combatants, or PSC); and principal amphibious ships (PAS).[10]

The personnel trend in Figure 11-3 shows a reduction of approximately one-third in regular naval personnel between 1998 and 2013, with further reductions to be implemented under the FF2020 plan. Although the number of PAS has been held constant over this period, the number of SSN has been reduced by more than 40 percent, and the number of PSC by almost 50 percent. The PSC trend line includes the temporary abandonment of UK aircraft carrier capability, but with at least one of the two Elizabeth-class carriers expected to be in commission by 2020.

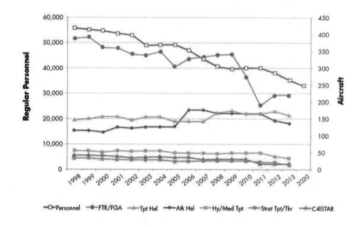

Source: "UK Defence Statistics Compendium," London, UK: UK Ministry of Defence, 1998-2013, February 25, 2014, available from *www.dasa.mod.uk/index.php/publications/UK-defence-statistics-compendium*; International Institute of Strategic Studies, *The Military Balance*.

Notes: The six aircraft categories are fighter and fighter/ground attack aircraft; attack helicopters; command, control, and communication aircraft; strategic transport/tanker aircraft; heavy/medium transport aircraft; and transport helicopters.

Figure 11-4. Deployable Air Forces, 1998-2013.

Figure 11-4 shows the regular personnel strength of the Royal Air Force and the number of aircraft in six key categories. Many UK combat aircraft (both fixed wing and rotary) are manned by personnel from both the Royal Navy and the British Army; these numbers are not represented in the shown personnel strengths.

Between 1998 and 2013, the personnel strength of the Royal Air Force shrunk by roughly 37 percent. In the same period, the number of fighter and fighter/ ground attack aircraft — arguably the most vivid and potent symbol of modern air power — were reduced by more than 40 percent. Yet the deployable strength of UK command, control, and communication aircraft (C4ISTAR) and attack helicopters and transport air- craft (fixed wing and rotary) — collectively essential for the projection of hard power in an expeditionary setting — were held more or less stable.

Taken together, Figures 11-2 to 11-4 clearly indi- cate major reductions in the military capability and personnel strength of UK armed forces since 1998. But the deeper significance of these changes is harder to gauge. Reductions in UK military power cannot be said to have been negligible, but neither do they seem to have been irreversible and fundamental.

Where national military posture is concerned, balance should be measured both quantitatively and qualitatively. Quantitatively, balance means consis- tency and equilibrium in political and budgetary com- mitment to land, sea, and air forces, respectively. By this definition, an unbalanced force would be one that sacrificed, say, air power to fund a naval construction program.

Qualitative balance is achieved by setting the size of a force, on the one hand, against its technological proficiency and military effectiveness, on the other.

Thus, if a smaller force with better equipment can achieve as much or more than a larger force with inferior equipment, then the smaller force might still be said to be balanced.

There are, of course, important gaps in this overview—most notably in aircraft carriers, carrier-borne fixed-wing air power, and C4ISTAR (including maritime patrol). Nonetheless, equipment programs are underway to help remedy these acknowledged deficiencies.

What should also be borne in mind in this survey of UK military capability is military equipment quality (MEQ).[11] For example, the MEQ of obsolescent aircraft such as the Jaguar cannot usefully be set against that of the Typhoon and the F-35 Lightning II; the Astute class of submarines is more capable than its predecessor, as is the Type 45 destroyer; and the A330 Voyager tanker aircraft will be more reliable than its predecessors. Old equipment is scarcely, if ever, replaced on a one-for-one basis; where military force is concerned, numbers and size are emphatically not everything.

It should also be borne in mind that "capability" has long since ceased to be synonymous with "weapon" or "weapon platform": modern military capability is best understood as a highly sophisticated, integrated C4ISTAR system. This is the case even at the level of the individual combatant. The modern infantry soldier, for example, should deploy on operations with a variety of high-quality personal, crew-served, and indirect-fire weaponry at his or her disposal. Body armor and vehicle protection have seen considerable improvements, while advanced communications, reconnaissance, and surveillance equipment have ensured unprecedented levels of battlefield situational awareness.

UK MILITARY OPERATIONS

UK armed forces approached the end of the 1990s having acquired considerable and varied operational experience. With the bulk of an armored division supported by air and sea power, the UK was the largest European contributor to the U.S.-led coalition operations in the 1991 Gulf War.

From 1992 to 1996, UK armed forces were closely involved in conflicts resulting from the breakup of Yugoslavia, contributing armed troops with air support to the "robust peacekeeping" mission of the United Nations Protection Force and the NATO-led Implementation Force. During the same period, the Royal Air Force contributed to NATO air campaigns in former Yugoslavia. It was not until the Good Friday Agreement in 1998 that the UK could begin to make substantial reductions in its considerable military commitment to Northern Ireland, where UK army, marine, and air force units had all acquired experience over decades in urban counterterrorism and counter-insurgency operations.

In December 1998, the UK and U.S. air forces undertook Operation DESERT FOX, a 4-day bombing campaign against targets in Iraq, in which Royal Air Force aircraft flew some 15 percent of the sorties. The following year, the UK air force participated in two NATO campaigns: Operation ALLIED FORCE against targets in the former Federal Republic of Yugoslavia and, later, Kosovo Force. That same year, one Royal Navy warship (HMS *Glasgow*) and a small contingent of British Army troops (with transport aircraft and helicopter support) participated in Operation WARDEN, the multinational peacekeeping force deployed to East Timor under Australian command.

In 2000, the UK mounted two joint operations involving land, sea, and air forces in Sierra Leone, Africa: Operation PALLISER to evacuate noncombatants from Freetown, and Operation BARRAS to rescue captured British troops. For some months during summer 2003, a small contingent of British land and air forces took part in Operation ARTEMIS, the European Union-led crisis management operation in the Democratic Republic of the Congo.

UK military involvement in Afghanistan began in late-2001 with a series of joint operations, including Operations VERITAS and FINGAL. In June 2002, Operation HERRICK, the UK contribution to the NATO-led International Security Assistance Force, was launched. Operation HERRICK has involved UK land, air, and sea forces in a series of 19 deployments. In 2009-10, at the height of the campaign, some 9,500 UK military personnel were deployed to Afghanistan, including infantry, light armor, artillery, and support troops, with fixed-wing and helicopter attack aircraft and transport and surveillance aircraft.[12]

In terms of scale, the most demanding UK military operation of the past decade was Operation TELIC, the British contribution to U.S.-led operations in Iraq in March 2003 and the subsequent civil-military occupation and counterinsurgency campaign that lasted until 2011. The UK deployed no fewer than 46,000 personnel at the start of the commitment, and the force comprised 30 navy warships and support ships, an armored division with three combat brigades and a logistics brigade, and the full range of fixed-wing and helicopter attack aircraft and transport aircraft.[13] As with Operation HERRICK, Operation TELIC made use of the roulement system, with 13 deployments between 2003 and 2011, each lasting about 5 to 6 months.

Finally, the UK also contributed to the military intervention in Libya from March to October 2011. Operation ELLAMY was principally a joint naval-air commitment: naval forces included principal surface combatants, cruise missile-firing submarines, mine countermeasure vessels, and a helicopter carrier (HMS *Ocean*); air forces included fighter, strike, C4ISTAR, and tanker aircraft, as well as attack and transport helicopters.

The UK armed forces have acquired a very high level of operational experience in the past decade and a half. British land, sea, and air forces have been involved in a wide variety of operations: from the very brief (4 days) to the very lengthy (13 years), from the relatively small (Operation WARDEN) to the very large (Operation TELIC), and in several different regions of the world (Africa, Europe, the Middle East, South Asia, and Southeast Asia). Although the roulement system used in Afghanistan and Iraq imposed significant strain on units and individuals, these long-term military operations in particular have spread the direct experience of warfare across UK armed forces.

UK STRATEGIC INTENT

UK national strategic outlook and intent are revealed in four sets of documents that punctuated its aforementioned operational experience: the SDR (1998), the SDR: *New Chapter* (2002), the *Defence White Paper 2003*, and the *National Security Strategy* and SDSR (2010).

1998: SDR.

The July 1998 SDR marked a fundamental transition in UK strategic outlook. The confusion and lack of direction brought about by the collapse of the 20th-century Cold War strategic order gave way to a mood of cautious engagement with an emergent 21st-century strategic order characterized by insecurity, diversity, and urgency.

The SDR had six significant themes, each of which has resonated powerfully in the UK national strategic debate ever since. The first of these was the Labour government's determination to conduct a "foreign-policy led strategic defence review" — an acknowledgement that in all the complexity of the emerging international security order, it made little sense for foreign policy and national defense to be considered separate domains.[14]

The second theme was risk management. The tone of the SDR was cautious:

> there is today no direct military threat to the United Kingdom or Western Europe. Nor do we foresee the re-emergence of such a threat. But we cannot take this for granted.[15]

Importantly, the argument advanced here was not that a defense posture of the Cold War style should therefore be maintained "just in case," but that it was a "vital British interest" that these benign trends should be encouraged by UK foreign policy. The implication for UK national strategy and defense was that they should move from "stability based on fear to stability based on the active management of these risks."[16]

The third theme of the SDR was affordability. The stated aim of the review was to "provide the country with modern, effective and affordable Armed Forces which meet today's challenges but are also flexible enough to adapt to change."[17] In an era where no "existential" threat to the UK could be identified, the defense budget would have to compete with other demands on public expenditure. In the uncertain times of the 21st century, spending on security and defense would be expected to provide a certain level of "future-proofing" in equipment acquisition.

Savings would also be achieved through technology, the fourth key theme, with the SDR calling for "much more precise application of force as a result of improvements in intelligence gathering, command and control and precision weapons." The fifth theme was alliance building: "For the foreseeable future we envisage that the largest operation we might have to undertake would be involvement in a major regional conflict, whether as part of NATO or a wider international coalition."[18]

The sixth and final theme of the SDR is captured in the term "expeditionary." The SDR promised "a fundamental reshaping of our armed forces" resulting in a "modernised, *rapidly deployable* and better supported front line."[19] Emphasis was laid on the effectiveness and efficiency of joint forces. The SDR confirmed the decision to build two new aircraft carriers with which UK maritime power would shift from "large-scale maritime warfare and open ocean operations in the North Atlantic" to "littoral operations and force projection."[20]

The SDR furthermore stressed the need to be able to deploy land forces, making use of improved battlefield reconnaissance and intelligence capabilities and new platforms such as the Apache attack helicopter. Air power, too, would be geared to expeditionary operations: the need for both air superiority and air defense would remain, but air defense of the UK would be at a lower priority.[21]

UK expeditionary force posture also shaped the SDR's defense planning assumptions (DPAs), according to which the UK would either "respond to a major international crisis" (such as a full-scale, tri-service commitment along the lines of the 1991 Gulf War) or "undertake a more extended overseas deployment on a lesser scale (as over the last few years in Bosnia) while retaining the ability to mount a second substantial deployment." In the event of the latter, dual commitment, the SDR would not expect "both deployments to involve warfighting or to maintain them simultaneously for longer than 6 months."[22]

2002: SDR: *New Chapter.*

No further, more formal account of the UK strategic outlook was published before the July 2002 appearance of the *New Chapter* to the SDR.[23] The *New Chapter* was an acknowledgement both of the events of September 11, 2001, and of the government's determination not to hold a formal national strategy and defense review so soon after the 1998 SDR. The new focus on terrorism as a strategic threat did, however, prompt an important change of emphasis in the DPAs:

> our analysis suggests that . . . several smaller scale operations are potentially more demanding than one or two more substantial operations. And there are now signs that frequent, smaller operations are becoming the pattern.[24]

2003: *Defence White Paper.*

The expeditionary theme, coupled more closely with the idea of long-distance counterterrorism and stabilization missions, was taken up again in the 2003 white paper, which was intended to build on both the SDR and the *New Chapter* "to provide a comprehensive statement of Defence Policy and an assessment of the strategic environment in which our Armed Forces operate." The *Defence White Paper* would be the "security and policy baseline against which future decisions will be made to enable the UK's Armed Forces to meet the full range of tasks they can expect to undertake in the future."[25] The document remained true to the expeditionary idea, albeit with a more pronounced counterterrorist flavor than the SDR:

> We must extend our ability to project force further afield than the SDR envisaged. In particular, the potential for instability and crises occurring across sub-Saharan Africa and South Asia, and the wider threat from international terrorism, will require us both to engage proactively in conflict prevention and be ready to contribute to short notice peace support and counter-terrorist operations.[26]

The *Defence White Paper* thus favored more, lighter, and smaller missions for the armed forces. This position was encapsulated in a subtle yet important shift in DPAs. Although the document insisted that "our forces must retain the capacity to undertake Large Scale operations at longer notice in Europe, the Mediterranean and the Gulf Region," the underlying UK strategic intention was clear enough: "Multiple concurrent Small to Medium Scale operations will be the most significant factor in our force planning." The

Defence White Paper continued, "We must therefore plan to support the three concurrent operations, of which one is an enduring peace support operation, that have become the norm in recent years."[27] This was no minor reorganization of existing military means: the 2003 *Defence White Paper* confirmed a significant change in strategic outlook as the UK began to focus more closely, and more explicitly, on "small wars."

2010: *National Security Strategy* and SDSR.

The most recent formal UK strategic review was published in 2010 in two parts: the *National Security Strategy* (NSS), published on October 18, and the SDSR, published the following day.[28]

The NSS was clear in one important respect: "we face no major state threat at present and no existential threat to our security, freedom or prosperity." Rather than thinking in terms of large-scale, traditional threats, the authors of the NSS thought in terms of risk: "The risk picture is likely to become increasingly diverse. No single risk will dominate." In a three-tiered list of "priority risks," the NSS set out a familiar mix of substrategic threats, in response to which UK conventional armed forces would most likely be used at long distance and at a relatively low scale. The four Tier One risks were international terrorism, cyber attacks and cyber crime, a major accident or natural hazard, and an "international military crisis between states" involving the UK and its allies.[29]

Other than in the case of the "international military crisis between states" (Tier One) and the case of the increased risk of terrorism resulting from "major instability, insurgency or civil war overseas" (Tier Two), the deployment of UK armed forces in the conventional role does not feature prominently in the first two tiers of the NSS priority risks table. Significantly, the possibility of a "major accident or natural hazard" appears as the third of four Tier One risks (a higher priority, therefore, than the "international military crisis"), while the prospect of a "large-scale conventional military attack on the UK by another state" appears only as a Tier Three risk.

To meet the wide and varied range of security risks and challenges set out in the NSS, the SDSR offered a new strategic policy framework that, in turn, generated eight national security tasks:

1. Identify and monitor national security risks and opportunities;

2. Tackle the root causes of instability;

3. Exert influence to exploit opportunities and manage risks;

4. Enforce domestic law and strengthen international norms to help tackle those who threaten the UK and its interests, including maintenance of underpinning technical expertise in key areas;

5. Protect the UK and its interests at home, at its border, and internationally to address physical and electronic threats from state and nonstate sources;

6. Help resolve conflicts and contribute to stability by, where necessary, intervening overseas, including legally using coercive force in support of vital UK interests, and protecting overseas territories and people;

7. Provide resilience for the UK by being prepared for all kinds of emergencies, being able to recover from shocks, and being able to maintain essential services; and,

8. Work in alliances and partnerships wherever possible to generate stronger responses.[30]

These eight tasks capture the wide-ranging and generally sober tone of both the NSS and SDSR. Yet they are not too narrowly concerned with hard power. Similarly, of the SDSR's 35 planning guidelines, only eight are directly concerned with the traditional, conventional use of military power. In keeping with the mood of caution and constraint, the SDSR's DPAs held to the pattern of the previous decade, expressing a preference for military deployments that would be either far away, fairly small, or relatively brief. The DPAs gave the following alternatives:

1. Conducting an enduring stabilization operation at around brigade level (up to 6,500 personnel) with maritime and air support as required, while also conducting one nonenduring complex intervention (up to 2,000 personnel) and one nonenduring simple intervention (up to 1,000 personnel);

2. Conducting three nonenduring operations if the UK is not already engaged in an enduring operation; or,

3. Committing, for a limited time and with sufficient warning, all UK military effort to a one-off intervention of up to three brigades, with maritime and air support (around 30,000, or two-thirds of the force deployed to Iraq in 2003).[31]

Judging by the 2010 NSS and SDSR, the UK strategic outlook is one in which the country will encounter a wide variety of security risks and challenges, ranging from natural hazards such as flooding to cyber crime to humanitarian crises to more traditional defense tasks, yet stopping short of an "existential" threat to the UK and its interests. Consequently, the armed forces are expected to undertake a wide variety of tasks, including early warning and intelligence gathering, aid to emergency organizations, the provision of a defense contribution to UK influence, and the projection of military power within the parameters of the 2010 DPAs.

ASSESSMENT

This chapter has charted the recent evolution of UK hard power in terms of four performance indicators. The first, military expenditure, has been relatively constant, while the second, military capabilities, shows downward trends, at least in quantitative terms. A qualitative assessment of UK hard power would certainly be more useful than a simple exercise in counting numbers. But because assessment methodologies are relatively underdeveloped, a qualitative assessment is beyond the scope of this chapter.

Operational experience, the third indicator of hard power, is the only one to show a firm upward trend: over the period covered by this chapter, and for several years previously, the UK acquired and consolidated a very strong reputation in the effective use of military force. The final indicator is strategic intent. Here, UK strategic rhetoric has very clearly shrunk.

I suggest three competing explanations for this mixed set of results. The first contender is that the evolution of UK hard power since 1998 has been driven largely by austerity, and remains so. By this view, the priority of successive governments has been to reduce the proportion of public expenditure devoted to defense as quickly as possible, accepting increased strategic risk in what is assumed to be a more benign world, to concentrate on restoring the health of the national economy.

The "peace dividend" of the early post-Cold War period, the argument might continue, was therefore no passing craze: it outlasted the 1990s and endured until it could be reincarnated in the post-2007 mood of austerity. There is a convincing aspect to this argument; UK defense is in the grip of austerity budgeting and will remain so for years to come, but it is also somewhat exaggerated. Since the 1990s, the UK government has, after all, spent a great deal of public money on defense: UK defense spending has consistently exceeded the NATO commitment to spend a minimum of 2 percent of GDP on defense, and the UK remains in the top ranks globally when it comes to defense spending.

The second possible explanation is that in recent years there has been a quiet campaign within government to design out UK capacity to act militarily. The purpose of this "anti-strategic" effort has allegedly been to dismantle UK hard power on the grounds that the capacity to intervene gave rise to the temptation to intervene, resulting in the deaths, injuries, expense, and reputational damage to the UK caused by Prime Minister Tony Blair-era wars, most notably in Afghanistan and Iraq.

Certainly, there are persistent, muttered allegations of government ministers and senior officials who have taken it upon themselves to exclude hard power from the UK strategic toolbox in preference for an emphasis on the so-called soft power of diplomacy, trading relations, and cultural interaction. By removing the capacity to act, the high-minded, internationalist, interventionist rhetoric of the government's declared strategic intent would thus become relatively free of risk and cost, since little could ever be done about it. Although national defense would still consume a large share of public spending, that sum would be far less than the cost of going to war.

This explanation is also unconvincing, however. Where defense, security, and military matters are concerned, soft power is valid and valuable, but it is best seen as a proxy for hard power rather than a sufficient alternative to it. It requires very little understanding of strategy to see that the result of a self-emasculation program would be for UK hard power to be replaced not by soft power but by bluff, and there might be nothing more expensive than the insecurity that is exposed when a bluff is called. Fortunately, it is barely conceivable that senior people charged with UK national security could have adopted such a strategically irresponsible, politically dishonest, and intellectually weak position.

If neither austerity nor anti-strategy offers a convincing explanation for the evolution of UK hard power, there is a third alternative. The only clear positive trend in the story of UK hard power over the past 15 or so years is the very high level of operational experience gained by UK armed forces. If the forces have remained so effective even under conditions of austerity, then it is at least possible that their success might have worked against them by providing a perverse disincentive for sustained investment in hard power, whether financial, intellectual, or political.

Operational experience might also mask the most convincing yet least attractive explanation for the current condition of UK hard power: strategic ambivalence. It cannot be said that the UK has altogether lost interest in hard power. But neither can it be said with much confidence what that interest is: Why should the UK remain interested in hard power? How important is hard power to the UK national strategic outlook? Is military expenditure seen as a government obligation or as a burden to be offloaded whenever and wherever possible?

Strategic ambivalence is a national strategic outlook that barely qualifies as such, where the aspiration is to commit as little as possible (politically and financially) while retaining the widest possible range of strategic options. Ambivalence can be seen at the political, strategic, financial, technological, and moral levels. Politically, the diminishing capacity for UK major operational deployments chimes with public antipathy toward large-scale military interventions, yet does not remove that option altogether.

Therefore, the expeditionary rhetoric found in the NSS and SDSR of 2010, as in earlier statements of strategic intent, need never be tested. Something of this sentiment can be found in a comment made by Secretary Hammond in oral evidence to the House of Commons Defence Select Committee in October 2013:

> It would be realistic of me to say that I would not expect — except in the most extreme circumstances — to see a manifestation of great appetite for plunging into another prolonged period of expeditionary engagement any time soon.[32]

Strategically, any adjustments in UK expeditionary hard power are offset by competence in other matters, such as counterterrorism and rescue operations, which are still perceived by the public and media to be serious national security tasks. Financially, the impressive reputation of UK armed forces holds out the alluring possibility that further cuts might be made (especially in personnel) without any obvious loss of competence, particularly if the scale and duration of operations are reduced.

Technologically, reductions in bulk hard power might rationalize a shift to a more technologically oriented posture involving intelligence, surveillance, precision strike, unmanned combat air vehicles, and similar features. These equipment and platforms are of interest not only because they are often less costly to operate than their conventional equivalents, but also because they offer a degree of political deniability that is not so readily available when there are boots on the ground. Technological warfare might even be considered morally preferable in that it should mean fewer troops being exposed to the risks of combat.

In some respects, strategic ambivalence is to be welcomed. At its most constructive, ambivalence could be the basis of a national strategy based on risk analysis and management—an approach that is most appropriate when national strategy must respond not only to a diverse range of security threats and challenges but also to scarcity and austerity.

Yet where matters of hard power are concerned, a national strategy based on ambivalence and risk must be deliberate rather than accidental and must involve careful and difficult decisions rather than expect to avoid them altogether. For a risk-based national strategy to be effective, it will require serious thought and investment in capabilities such as intelligence gathering, early warning, and communications.[33] It remains to be seen whether the UK government will remain meaningfully committed to a risk-based national strategy.

In his first speech as chief of the defense staff in December 2013, General Sir Nick Houghton observed that "[UK] Defence has for many years, certainly since the end of the Cold War, and in strong international company within Europe, been managing the decline of military hard power."[34] But managed decline is not the same as decline; there must be strategic capacity and purpose in whatever remains of the process — however inevitable—of retrenchment.

As the basis for national strategy, ambivalence is no substitute for analysis and decision, and it cannot offer a reassuring glimpse of the future; national strategy will continue to require complex judgments that are periodically revised as circumstances change. Finally, it is unwise to expect to be ambivalent about everything in national strategy, particularly hard power: national hard power either exists on a militarily meaningful scale or it does not; it either has purpose or it does not.

ENDNOTES - CHAPTER 11

1. This chapter was originally published as an essay on July 8, 2014.

2. *Strategic Defence Review*, London, UK: UK Ministry of Defence, July 1998, p. 5, available from *fissilematerials.org/library/mod98.pdf*.

3. James Blitz, "Fox Claims 'Black Hole' of Defence Costs Eliminated," *Financial Times*, September 26, 2011, available from *www.ft.com/cms/s/0/22559c66-e847-11e0-9fc7-00144feab49a.html#axzz3325j39Yh*; and James Blitz, "MoD Plugs £38bn Budget 'Black Hole'," *Financial Times*, May 14, 2012, available from *www.ft.com /cms/s/0/efed7ef2-9de9-11e1-9a9e-00144feabdc0 .html#axzz3325j39Yh*.

4. *UK Ministry of Defence, MoD Reveals £160 Billion Plan to Equip Armed Forces*, January 31, 2013, available from *www.gov.uk/government/news/mod-reveals-160-billion-plan-to-equip-armed-forces*.

5. Paul Cornish and Andrew Dorman, "Fifty Shades of Purple? A Risk-Sharing Approach to the 2015 Strategic Defence and Security Review," *International Affairs*, Vol. 89, No. 5, September 2013, pp. 1183–1202, available from *www.chathamhouse.org/publications/ia/archive/view/194088*.

6. Ben Farmer, "Wars in Iraq and Afghanistan Were a 'Failure' Costing £29bn," *The Telegraph*, May 28, 2014, available from *www.telegraph.co.uk/news/uknews/defence/10859545/Wars-in-Iraq-and-Afghanistan-were-a-failure-costing-29bn.html*.

7. Cornish and Dorman, "Fifty Shades of Purple," p. 1187.

8. All personnel data in Figures 11-2 to 11-4 have been drawn from official UK publications. See *UK Defence Statistics Compendium (1998–2002)* and *UK Defence Statistics Factsheets (2003–13)*, London, UK: UK Ministry of Defence, available from *www.dasa.mod.uk/index.php/publications/UK-defence-statistics-compendium*. Since a single, consistent data series on UK armed forces personnel strength is not readily available, personnel data entries are best read as trend indicators rather than real values.

9. For the purposes of this chapter, land-force personnel include the regular, full-time strength of the British Army (as published) plus 6,000 to cover the nominal strength of the Royal Marines.

10. For the purposes of this chapter, naval personnel include the regular, full-time strength of the Royal Navy (as published) minus 6,000, representing the nominal strength of the Royal Marines counted in Figure 11-2.

11. For leading analysis on this subject, see Steven Bowns and Scott Gebicke, "From R&D Investment to Fighting Power, 25 Years Later," *McKinsey on Government*, No. 5, Spring 2010, pp. 70–75, available from *www.technology-futures.co.uk/MoG5_DefenseR&D_VF.pdf*.

12. International Institute for Strategic Studies, "Chapter Four: Europe," *The Military Balance 2011*, Vol. 111, No. 1, 2011, p. 160; and "Chapter Four: Europe," *The Military Balance 2012*, Vol. 112, No. 1, 2012, p. 172.

13. BBC News Middle East, "Iraq War in Figures," December 14, 2011, available from *www.bbc.co.uk/news/world-middle-east-11107739*.

14. *Strategic Defence Review*, London, UK: UK Ministry of Defence, p. 8.

15. *Ibid.*

16. *Ibid.*

17. *Ibid.*, p. 9.

18. *Ibid.*, pp. 13, 16.

19. *Ibid.*, p. 5 (emphasis added).

20. *Ibid.*, p. 30.

21. *Ibid.*, p. 25.

22. *Ibid.*, p. 26.

23. See *The Strategic Defence Review: A New Chapter*, London, UK: UK Ministry of Defence, July 2002, available from *www.comw.org/rma/fulltext/0207sdrvol1.pdf.*

24. *Ibid.*, p. 14.

25. Policy Director, UK Ministry of Defence, *The Defence White Paper & Operations in Iraq – Lessons for the Future*, London, UK: UK Ministry of Defence, December 11, 2003, available from *webcache.googleusercontent.com/search?q=cache:http://webarchive.national archives.gov.uk/%2B/http:/www.mod.uk/publications/iraq_future lessons/chap3.htm.*

26. *Delivering Security in a Changing World: Defence White Paper*, London, UK: UK Ministry of Defence, December 2003, p. 7, available from *www.mocr.army.cz/images/Bilakniha/ZSD/UK%20 Defence%20White%20Paper%202003.pdf.*

27. *Ibid.*, p. 7.

28. "Her Majesty's Government, Securing Britain in an Age of Uncertainty," *The Strategic Defence and Security Review*, October 2010, available from *www.gov.uk/government/uploads/system/up-loads/attachment_data/file/62482/strategic-defence-security-review.pdf.*

29. All quotes in this paragraph can be found at *A Strong Britain in an Age of Uncertainty: The National Security Strategy*, London, UK: Her Majesty's Government, pp. 15, 18, 27, available from *www.gov.uk/government/uploads/system/uploads/attachment_data/ file/61936/national-security-strategy.pdf.*

30. *The Strategic Defence and Security Review*, Her Majesty's Government, pp. 10–12.

31. *Ibid.*, p. 19.

32. House of Commons Defence Committee, *Towards the Next Defence and Security Review: Part One*, January 7, 2014, available from *www.publications.parliament.uk/pa/cm201314/cmselect/ cmdfence/197/197.pdf.*

33. Paul Cornish, *Strategy in Austerity: The Security and Defence of the United Kingdom*, London, UK: Chatham House, 2010, pp. 23–24, available from *www.chathamhouse.org/publications/papers/ view/109490.*

34. General Sir Nicholas Houghton, "Annual Chief of the Defence Staff Lecture 2013," lecture, RUSI, London, UK, December 18, 2013, available from *www.rusi.org/events/past/ ref:E5284A3D06EFFD.*

CHAPTER 12

POOLING AND SHARING:
THE EFFORT TO ENHANCE ALLIED DEFENSE
CAPABILITIES[1]

W. Bruce Weinrod

KEY POINTS

- In the face of declining defense budgets, transatlantic allies have shown increased interest in pooling-and-sharing defense efforts.
- The results have been mixed, with some notable successes, such as the Strategic Airlift Capability (SAC), and other less-successful efforts, such as the European Union's (EU) attempt to establish a common training program for jet pilots.
- Large-scale pooling and sharing tends to infringe on national sovereignty issues. Consequently, successful programs are most likely tied to discrete areas of cooperation and are often carried out by smaller groups of nations.

Recent developments in Crimea and Ukraine highlight the crucial importance of robust transatlantic military capabilities. However, these capabilities are on a downward trajectory. If current trends continue, the weakening of collective defenses may reach a tipping point where significant collective power projection would be problematic at best.

This need not happen. The transatlantic nations have the collective resources to ensure a credible and robust defense capability, and allied governments continue to proclaim the need for a capable national defense while believing that their own security can best be ensured by joining in common defense commitments and programs.

To these ends, transatlantic defense officials are giving increased attention to ways in which defense budgets can be more efficiently and effectively allocated. Most allies are largely rejecting budget increases as unfeasible in today's economic climate and are looking to each other to better utilize existing resources through pooling-and-sharing efforts.

This exploration is all the more urgent, given that the United States may not always provide substantially more than its fair share of transatlantic military resources. (According to most recent figures, the U.S. share of allied defense expenditures was over 70 percent.[2]) The September 2014 NATO summit in Wales, United Kingdom (UK), presents the most promising overall opportunity for the highest levels of allied governments to provide an impetus for the full development and implementation of pooling-and-sharing initiatives.

BACKGROUND

"Pooling" and "sharing" are complementary terms applied to various cooperative defense arrangements for bringing together the resources of two or more nations to enhance effectiveness or lessen costs. Pooling and sharing can be accomplished within a transatlantic or European-wide framework (such as the North Atlantic Treaty Organization [NATO] or the EU) or through dedicated bilateral or multilateral arrangements.

Although pooling-and-sharing defense programs have attracted substantial attention in recent years, such arrangements existed before the emergence of the term. What is new is the high priority being placed on pooling-and-sharing projects as a way to reduce costs to individual nations, while ensuring the existence of necessary military capacities.

At the same time, implementing pooling-and-sharing programs on a large scale is challenging. For example, programs tied to logistics or to tactical intelligence are relatively easy to implement. However, activities that delve more deeply into operational capabilities can become more politically and economically complex. Further, nations considering pooling and sharing can face difficult decisions regarding the allocation of defense resources that impact national defense industries and that rely on other nations to provide necessary military capabilities in times of crisis or conflict.

NATO POOLING AND SHARING

Pooling and sharing is a priority objective for NATO and is included in its broad Smart Defence initiative, which was launched by current and outgoing NATO Secretary General Anders Fogh Rasmussen in 2011 and was affirmed at the 2012 NATO summit.[3] Pooling and sharing has a number of antecedents within NATO.[4] The best known is the NATO airborne early warning and control system (AWACS), which became operational in December 1978. The force consists of 17 E-3A aircraft and is supported by 18 participating NATO nations, which share operational costs. The UK makes an in-kind contribution of its E-3D aircraft. The AWACS has proven to be a successful, cooperative program that has provided an important operational capability, including, most recently, its deployment in the Afghan theater of operations.

Separately, NATO operates a jet-fuel pipeline linking 13 NATO nations to provide for NATO requirements. NATO also utilizes the NATO Support Agency, which handles the organization's logistics and procurement. The pipeline and support agency are examples of pooling-and-sharing initiatives launched in the early days of the alliance.

A more recent cooperative program that fulfills a key operational need for participating nations is the SAC, a 12-member consortium that, at present, deploys three C-17 transport aircraft. SAC aircraft are available to contributing nations to meet their national military needs, including those related to NATO and European Union (EU) commitments. SAC consortium nations include 10 NATO nations and NATO Partnership for Peace members, Finland and Sweden. Based in Hungary, SAC aircraft utilize multinational crews and are supported by personnel from all participating nations.

SAC members need the lift capability of large aircraft, but most do not have the financial resources to acquire or maintain such a major capability. Thus, the SAC offers a cost-effective approach that permits participants to purchase specific sets of flying hours as needed. All SAC partners pay operational costs and can utilize the SAC for any purpose such as airdrops and assault landings. The consortium has already been used for operations in the Balkans, Afghanistan, and Libya, as well as for peacekeeping and humanitarian relief operations.

The SAC also provides an alternative structural model to the AWACS. While AWACS aircraft are owned by NATO and are thus part of NATO's overall structure, SAC aircraft are owned by a legally separate consortium of nations that includes both NATO and non-NATO countries and that has an arrangement that allows the SAC to use NATO support structures. One advantage of SAC arrangements is that, while the AWACS program requires unanimous consent by its partners for use, the SAC does not have such a requirement.

NATO is also currently developing another commonly supported capability known as the Alliance Ground Surveillance (AGS) system, which is to include five Global Hawk high-altitude unmanned aerial vehicles (UAVs), with operations expected to begin in several years. These UAVs will be deployed with sophisticated radars and will allow NATO to monitor ground activities over wide areas and under all weather conditions.

The AGS will be purchased by 14 NATO nations. Its infrastructure and operational support will be funded through NATO's common funding program, which is composed of financial contributions from all NATO members. In addition, France and the UK will provide in-kind service support, while other nations will supplement the AGS with national air surveillance capabilities as required. Industries from all of the AGS nations will participate in the AGS program, and all NATO nations will have access to AGS-acquired information. As with the AWACS, AGS will be a NATO system with the international status of a formal subsidiary organization of NATO. Program management will be provided by the NATO Alliance Ground Surveillance Management Agency.

The aforementioned programs reflect the fact that pooling-and-sharing activities can involve establishing a dedicated coordinating framework. Indeed, NATO has established specific bureaucratic structures for the AWACS and SAC. Other dedicated bureaucratic structures that were launched in the early days of the alliance, such as the NATO pipeline and NATO Maintenance and Supply Agency, are also examples of successful pooling-and-sharing programs.

There are also a number of more recently established NATO pooling-and-sharing arrangements. For example, multiple NATO nations take turns providing fighter aircraft to patrol the airspace over the three Baltic allied states of Estonia, Latvia, and Lithuania. As a result, these smaller nations do not need to acquire this capability and can spend their defense resources on other priorities. Similarly, Germany provides maritime surveillance in the North Sea, thus alleviating the need for such a capability on the part of the Netherlands.

Near-term projects envisioned by NATO include the development of a multinational cyber defense capability; creation of a multinational chemical, biological, radiological, and nuclear battalion with pooled equipment and training; establishment of a multinational aviation training center for helicopter pilots and ground crews; and pooling of allied maritime patrol aircraft.

EU POOLING AND SHARING

Just as NATO has capitalized on the pooling and sharing of resources, the EU has adopted pooling and sharing as a focal point for efforts to develop common European security programs. In 2004, the EU established the European Defence Agency (EDA) as a framework for coordinating European defense cooperation activities.[5] EU-EDA pooling-and-sharing efforts have consisted of a modest number of specific projects and a variety of planned initiatives.

An important ongoing program area facilitated by the EDA involves military air transport. In 2010, the EDA began operation of the European Air Transport Command (EATC), based in the Netherlands. The EATC coordinates military transport fleets of its five member nations (France, Germany, the Netherlands, Belgium, and Luxembourg) and undertakes occasional exercises and training programs. In 2013, eight nations—four EATC states plus the Czech Republic, Italy, Sweden, and Spain, the host nation—participated in the European Air Transport Training Exercise.

In addition, the EDA has established the framework for a European Air Transport Fleet (EATF). The EATF has 20 members but is currently more of a notional structure than an operational enterprise. Over time, the EDA would like the EATF to expand to include the exchange or acquisition of aircraft and supporting capacities, including maintenance, cargo handling, and common training.[6]

An important recent development was the Ghent Initiative, presented to the EU in September 2010 by Sweden and Germany. The initiative proposed that the EDA could, in close cooperation with other organizations, coordinate and potentially link various EU pooling-and-sharing efforts. The initiative also urged EU nations to divide their military capabilities into several categories: capabilities that are indispensable to the state's security and need to be maintained exclusively by the state, capabilities that could be maintained in closer cooperation with partners without the state losing authority over them (pooling), and capabilities that could be eliminated when provided by other states (sharing).

Over the past few years, EU member states have put forward many ideas for enhanced pooling and sharing in areas that are key military and operational enablers: strategic transport, air-to-air refueling, medical support, surveillance and reconnaissance, maritime surveillance, pilot training, naval logistics, and military communication satellites. As yet, however, there has been halting progress in implementing these ideas at the EU level.[7]

Independently, the EDA has offered to assist EU nations with pooling and sharing in areas such as shared use of fixed military infrastructure and in defense acquisition and manufacturing. The EDA is also assessing potential projects in maritime surveillance capabilities, surveillance and reconnaissance, military satellite communications, smart munitions, and naval logistics. Further, in November 2012, the EDA promulgated a voluntary code of conduct whose stated purpose is to support cooperative efforts to develop defense capabilities.

That said, a substantial part of EDA activities have thus far consisted of studies and recommendations as opposed to actual programs. The absence of more definitive defense projects is a result of several factors: 1) the lack of major financial commitments for such projects, 2) the view in most EU nations that national interests, including sustaining national defense industrial bases-take priority over cooperative endeavors, and 3) the firm UK position that EDA programs must be limited in scope and cost.

BILATERAL AND MULTILATERAL EFFORTS

In addition to pooling-and-sharing activities within NATO and EU frameworks, various European nations have developed bilateral or multilateral defense relationships that include pooling and sharing. Such arrangements typically involve nations in geographical proximity. The following subsections detail some of these arrangements.

Nordic Defence Cooperation.

Established in 2009, the Nordic Defence Cooperation (NORDEFCO) framework has reinforced an already-existing history of cooperation among the Nordic countries of Denmark, Norway, Finland, Sweden, and Iceland. Defense cooperation has focused on joint training, exercises, and capability development.

Indeed, NORDEFCO holds promise to be the most advanced regional grouping in the years ahead. At a December 2013 meeting, NORDEFCO nations agreed on a future plan of action, outlined in the *Nordic Defence Cooperation 2020*.[8] The plan calls for air surveillance patrols (by Sweden, Finland, and NATO members, Norway and Denmark) over Iceland.

NORDEFCO also announced its Cooperation Air Transportation initiative for using air transport assets and, possibly, in the future, pooled efforts in the areas of maintenance, spare parts, and procurement. Nordic countries will also, according to the December 2013 plan, focus on developing joint rapid deployment capabilities, including Arctic missions, along with developing new rules and processes for enhancing prospects for joint procurements.

The Visegrad Group.

The Visegrad Group (VG), established in 1991, consists of the Central European nations of Poland, Hungary, the Czech Republic, and Slovakia. The VG was formed to increase cooperation among the four countries in a range of policy areas, including defense.

Early VG efforts to enhance collaboration on defense projects were unsuccessful. More recently, catalyzed by the EU Ghent Initiative and the NATO Smart Defence program, the VG has given increased attention to cooperation in this area. One tangible result has been an agreement to develop the Visegrad Battle Group, with an operational target date in the first half of 2016.[9] The group is projected to consist of approximately 3,000 troops, and Poland will serve as its lead nation. In addition, the VG has helped coordinate training for helicopter pilots under the NATO HIP helicopter support program.[10] Separately, the Czech Republic and Slovakia (both VG nations) have joined with the United States and Croatia to develop a Multinational Aviation Training Centre to train crews of Russian-made, Mi-8-type helicopters.[11]

The Weimar Triangle.

The Weimar Triangle (WT) was established in 1991 as a mechanism for cooperation among France, Germany, and Poland. As with the VG, the WT has focused principally on political, economic, and cultural relationships. Defense efforts have consisted mainly of meetings and communiqués. However, as with the VG, the WT agreed to develop a Weimar Triangle Battlegroup consisting of 1,500 troops ready for rapid deployment, which became operational in 2013.[12]

The France-UK Defense Treaty.

Seeking to work around constrained defense budgets, in 2010, the UK and France reached an agreement envisioning significant pooling-and-sharing efforts in which the two nations would share nuclear-weapons testing facilities, defense research, and aircraft carriers. The agreement also called for developing a joint expeditionary force and for cooperation in maintenance, training, and logistics in connection with the French and British air forces' acquisition of the A400M transport aircraft.[13]

Implementation of the agreement has been thus far incomplete but has included regular joint military exercises. Some British officers have been deployed on a French aircraft carrier,[14] and pilots from each country have flown the other's fighter jets as an initial step toward establishing a combined joint expeditionary force by 2016.[15] France has also agreed to a British proposal to jointly develop and build a new anti-ship missile and a new generation of advanced unmanned aircraft.[16]

Benelux Defense Cooperation.

Defense cooperation between Belgian and Dutch naval forces began in the early post-World War II period and has included the establishment of a single commanding officer for the two navies, of an integrated naval staff, and of integrated support structures.[17] An accord was signed in 2012 for increased joint naval training between Belgian commandos and the Dutch Airmobile Brigade. The two air forces also agreed to cooperate more closely in using each other's airfields, in joint deployments, and in integration of materiel support. Future areas of cooperation include logistics and maintenance, military education, defense acquisition, and joint military operations.[18]

Another multilateral arrangement involving Belgium and the Netherlands is the European Participating Air Forces (EPAF) program, which also includes

Denmark and Norway. EPAF emerged from the initial acquisition of F-16s by these nations. In the ensuing years, the nations have trained together and used common logistics facilities. Although not formally a part of NATO, EPAF fighters have deployed on NATO missions in the Balkans and Afghanistan.[19]

KEY CHALLENGES FOR POOLING AND SHARING

Pooling and sharing can be an important mechanism for maintaining transatlantic military capabilities over the longer term. At the same time, economic, technological, military, and political challenges exist. Any of these challenges, much less a combination of them, can make the successful development and implementation of pooling-and-sharing projects difficult to accomplish.[20]

Economic issues can delay or even block pooling-and-sharing programs. Domestic-based defense industries can place great pressure on their respective governments to gain the largest possible share of work in any cooperative project. Resolving the conflicting interests may take lengthy negotiations that can substantially delay, or even render futile, the development of a new initiative.

For example, the AGS program discussed earlier was first proposed by NATO in the late-1980s. Despite AGS having been the top priority for NATO military leaders, it took a decade and a half for NATO leaders to reach the agreement to deploy AGS, with the delay due largely to discussions about which defense companies in which countries would get what share of the work.

In addition, nations working together do not always produce a more cost-effective outcome or a more effective capability. With shrinking defense budgets, there is even more pressure to distribute development dollars and acquisition dollars to keep companies afloat. In some instances, this can lead to the division of project work in inefficient or not-cost-effective ways, including requirements that a certain percentage of work be allocated to specific nations to assuage domestic constituencies.

For example, the NH90 helicopter, developed under a multilateral program by NATO nations, witnessed significant delays and a series of technical problems in development, with further complications arising from the fact that the helicopter was being designed for use by different military services and had different design configurations for some nations. The result was a significant increase in expected costs— probably well above what the system would have cost if the helicopter had been developed by one nation. A similar story can be told regarding the multinational programs to develop and build the Eurofighter Typhoon and A400M transport plane.

Coordination of pooling-and-sharing projects among participants can also prove challenging. Nations have different planning, programming, and budgeting cycles. For example, budget cycle variances were the principal cause of France's hesitation and delays in agreeing to a UK request to develop jointly an anti-ship missile system. Moreover, nations participating in pooling-and-sharing efforts need to reach agreements on such matters as system ownership, military command structures, the use of bases, and possible national caveats regarding the conduct of actual military operations.

The proposed Advanced European Jet Pilot Training System illustrates some of these difficulties. The program, whose planning began a decade ago and which was approved by the EDA in 2009, contemplated using European bases for pilots to train on a common fleet of aircraft. However, national differences about key elements of the project—including which bases and aircraft should be used—and about training methods have prevented the program from moving forward. Interestingly, a similar program already exists within NATO, which for many years has conducted jet pilot training for 13 nations at Sheppard Air Force Base, Texas.

Domestic considerations can also delay or block the implementation of a project, even after a government agrees to join a specific program.[21] Finance ministries or individual military services may balk at providing the funding necessary to implement an agreement. For example, the United States pledged at an early point to participate in the NATO SAC program, but it delayed signing the required memorandum of understanding to begin American participation because the U.S. Air Force was reluctant to provide the necessary funding from its budget. It finally did, but only after being ordered to do so.[22]

It must also be kept in mind that both NATO and the EU can only undertake those projects that member nations authorize and are willing to fund. In addition, there is at present an inherent limit on EDA activities, given the differing EU member views on the nature and extent of the EU's role in security matters.

Pooling and sharing requires the cooperation of national militaries, and while multinational coordination has occurred successfully at times, such arrangements face hurdles in overcoming differences in capabilities, in military doctrines, in weapons systems, in training, and in personnel and command structures.

In addition, pooling and sharing has the potential to constrain collective military capabilities. If, for example, a nation entirely gives up a certain capability in deference to another nation with similar capabilities, the loss of redundancy could be a problem in protracted or large-scale conflict operations. Also, interoperability gaps may make it challenging for a nation with advanced capabilities to work effectively with a smaller nation that possesses a similar but less-developed capability.

Of course, a critical concern is whether shared capabilities will be available when needed. If a nation possessing a necessary capability refuses to participate in a collective military action, this will obviously make it more difficult to carry out an operation and could even alter the calculus of whether to undertake the operation altogether. Furthermore, a fundamental concern standing in the way of large-scale pooling and sharing is how it infringes on national sovereignty. By eliminating a particular capability, it is argued that a nation becomes dependent on other nations to provide capabilities necessary for its own security and, in turn, risks its ability to carry out a core task of a nation-state.

MAKING POOLING AND SHARING WORK

Even considering the limitations and challenges outlined previously, pooling and sharing has the potential to be an important instrument for helping sustain necessary transatlantic military capabilities, especially in the current constrained economic environment. Pooling-and-sharing efforts can be most successful if they follow the guidelines outlined in the following list.

1. NATO should be open to creative pooling-and-sharing arrangements, such as the SAC, which includes NATO (and non-NATO) members but is not within NATO.

2. Although rarely popular in member states' capitals, common funding (for example, NATO nations' financial contributions for common endeavors) is necessary to carry out new programs.

3. NATO should reform its decisionmaking processes (especially those NATO procedures that currently call for unanimity on all major decisions) to ensure that one nation cannot block or delay a project indefinitely.

4. NATO should explore pooling-and-sharing arrangements that are developing under the recently proposed initiative to establish "framework nations." Under this approach, member states who have retained a broad range of military capabilities would act as lead nations in coordinating programs with an eye to meeting NATO defense planning targets on a tailor-made, multinational, but not alliance-wide basis. Smaller allied militaries would then plug into the enabling capabilities that only the big nations can provide (for example, air-to-air refueling or strategic surveillance and reconnaissance).

5. Nations should be more forthcoming with defense plans. So far, the practice has been for individual member states to consider, decide, and announce defense cutbacks without either consulting with or

informing allies or NATO. Thus, under current circumstances, important NATO capabilities can be weakened without any advance opportunity to consider how they might be maintained in different ways.

Such advance notice can allow time for an assessment to be made regarding the nature and extent of any effect of a reduction in collective transatlantic military capabilities. A dedicated evaluation center, such as a consortium of independent policy organizations, should be established. This center could assess national plans and their effect on collective capacities and could suggest alternative approaches — including pooling and sharing — that might maintain or develop a needed capability.

6. It bears repeating that the EU and NATO must avoid redundancy. There are simply insufficient resources for the organizations to be undertaking duplicative projects. At the same time, both organizations should strive to identify synergies.

In theory, both organizations recognize the need for enhanced coordination. The 2012 NATO summit specifically endorsed European programs to strengthen air-to-air refueling capacities. Further, NATO's Allied Command Transformation (ACT) has developed a number of potential pooling-and-sharing projects that consult with the EDA. In addition, NATO and the EU meet regularly in the EU-NATO Capability Group to discuss common capability requirements. NATO's ACT and the EU's EDA are also in regular contact.[23]

There may also be programs that NATO chooses not to pursue that would nonetheless enhance overall transatlantic security or provide capabilities helpful for localized contingencies. In this regard, the EDA Code of Conduct's call for giving pooling-and-sharing programs priority protection from defense cuts is sensible, and the permanent, structured cooperation arrangements envisioned under the EU Treaty of Lisbon could provide a mechanism for pooling and sharing by small groups of nations.[24]

7. Pooling and sharing should focus on practical cooperation that can truly enhance capabilities. Such cooperation also has the added virtue of often being the easiest to implement. Indeed, as described earlier, there are already a number of ongoing pooling-and-sharing arrangements in basic areas such as education, training, and exercises. More advanced efforts could focus on developing multinational logistics-and-maintenance support for selected capabilities. In any event, NATO can and should build on the cooperative efforts developed during its mission in Afghanistan.[25]

It could also make sense for multilateral programs to address core mission or functional areas, as the EATC does for airlift and air-to-air refueling. One approach would be to cluster nations by mission area (such as maritime power projection), function (airlift), or system (for example, NH90 helicopter users or F-35 users).[26]

8. Industry has an essential role in ensuring that pooling-and-sharing projects are cost effective and of requisite quality. To facilitate that goal, a focus on industrial issues must begin at the conceptual stage of a project to work out arrangements that ensure positive industry participation that continues throughout the program's operations or life cycle.[27]

External factors—including a more open European defense market, more cross-border cooperation, or mergers among European defense companies—can also facilitate industry involvement in pooling and sharing. Since most of the government-to-government frameworks for international defense cooperation have thus far been bilateral, the regulations and procedures for armaments cooperation need to be adapted for multinational procurement and cooperation.

9. Enhanced coordination of national defense programing timetables is also essential for significant pooling and sharing to succeed. Nations must exchange information on projected national procurement processes to ensure that production cycles in such areas as requirements, development, procurement, and maintenance are in sync. All such coordination should also be linked to NATO defense planning, as appropriate.

Further, the number of variants of military systems should be minimized. This will make maintenance and training, coordination of doctrine and operational concepts, and the use of common logistics easier and less costly. Of course, the most effective allied cooperation requires that military systems be interoperable, with common standards and certifications.

10. Bottom-up involvement in pooling and sharing is likely to be the most successful approach. As described previously, pooling-and-sharing activities by small groups of nations have been ongoing, for decades in some instances, and such regional and subregional arrangements will likely be the major engine for more of these arrangements. Experience indicates that projects organized by nations that are geographically close, that generally share common values, and that have similar threat perceptions are more likely to be developed and actually implemented. At the

same time, such groups can take in additional participants when practical and when likely to enhance capabilities.[28]

However, a challenge for such bottom-up initiatives is to ensure coordination with NATO. A multiplication of separate multinational arrangements might not enhance overall transatlantic military capabilities, so these efforts should be coordinated within a NATO framework to ensure maximum results.

11. A key aspect of 21st-century NATO that is growing in importance is NATO's partnership structure. Non-NATO nations have in recent years assumed an increasing role in NATO programs and activities. Thus, it is essential to enhance NATO's relationship with its most militarily capable partner nations and to identify further pooling-and-sharing programs in which such partners can participate. Countries such as Finland and Sweden bring much to the table and are already working closely with NATO.[29]

12. Finally, and most crucially, concerns about the potential availability of military assets during a time of crisis or conflict understandably exist in a number of nations. Unless or until this issue is resolved, it will likely place inherent limits on the nature and extent of pooling and sharing and, thus, pooling and sharing will of necessity need to focus on discrete and manageable capabilities.

CONCLUSION

The protection of U.S. and transatlantic national security interests and, indeed, the furtherance of crucial foreign policy objectives in general cannot be successfully managed without underlying credible and robust military capability.[30] While specific required military capabilities may be different than those of the past, the need for a highly capable and ready military remains. The Russian invasion of Ukraine and annexation of Crimea is a useful reminder that military force remains a fact of international life—if a reminder was even needed.

Ensuring necessary defense capabilities has become much more challenging because of the economic downturn and the absence of a shared consensus about the threats the democratic West faces. As a result, European military capabilities are without a doubt in decline. However, collective European military resources can still produce capable military forces if exploited in an effective manner. Such forces could handle security issues within the European region and adjacent areas, with the United States cooperating as appropriate, and be in a position to participate in international security coalitions addressing broader threats involving larger-scale cooperation between the United States and transatlantic nations.

The September 2014 NATO summit should endorse a specific pathway for the development of a robust transatlantic pooling-and-sharing program. NATO and the EU can both provide structural frameworks for pooling-and-sharing activities, but NATO can and should remain the principal mechanism for transatlantic military cooperation. NATO remains the strongest global military organization and is also the institutional link connecting the United States directly to transatlantic security.

That said, as previously noted, pooling and sharing need not always emanate from NATO or the EU. Various regional groupings are already playing a role in enhancing cooperation, and these organizations may well prove to be more effective in this regard than either NATO or the EU. But this should be accomplished within broader institutional frameworks and in a consistent manner with broader alliance capability requirements.

At the end of the day, it is most important that the requisite military capabilities exist and are available when needed to protect transatlantic security interests. If nations have the political will to allocate necessary funds and address coordination issues, pooling and sharing can be a key mechanism for the development of necessary technologies and weapons systems and can play an important role in the maintenance of essential transatlantic military capabilities.

ENDNOTES - CHAPTER 12

1. This chapter was originally published as an essay on August 11, 2014.

2. North Atlantic Treaty Organization, Secretary General's Annual Report 2013, January 27, 2014, available from *www.nato.int/cps/en/natolive/opinions_106247.htm*.

3. NATO describes Smart Defence as "a renewed culture of cooperation that encourages allies to cooperate in developing, acquiring, and maintaining military capabilities to undertake the Alliance's essential core tasks. . . . That means pooling and sharing capabilities, setting priorities and coordinating efforts better." For details on the Smart Defence initiative, see North Atlantic Trade Organization, "Smart Defence," available from *www.nato.int/cps/en/natolive/78125.htm*.

4. For example, in 2002, NATO's Prague Capabilities Commitment called for "multinational commitments and pooling of funds." For more information on the Prague Capabilities Commitment, see Carl Ek, "NATO's Prague Capabilities Commit-

ment," January 24, 2007, available from *www.fas.org/sgp/crs/row/RS21659.pdf*.

5. It should be noted as context that within the EU, there are differences regarding the nature and extent of European-only defense activities. The UK, in particular, has voiced a string of concerns about any EU defense projects or capabilities that would undermine the primacy of NATO or its own capabilities. That said, there is general acceptance by EU members of the concept of pooling and sharing as an appropriate EU activity.

6. The EDA envisions that the EATF will consist of a framework federating different projects identified, different structures and different types of assets, in order to create synergies through far-reaching cooperation and coordination. It will be a flexible and inclusive partnership between national and multinational military air transport fleets and organizations in Europe, aimed at the enhancement of standardized air transport services through cost-effective pooling, sharing, exchange and/or acquisition of various capabilities, including aircraft, training programs, cross-servicing activities, cargo handling, maintenance activities, spare parts, etc. See European Defence Agency, European Air Transport Fleet (EATF) Fact Sheet, May 19, 2011, available from *www.eda.europa.eu/docs/documents/factsheet_-EATF_final*.

7. See *Joint Statement on the Common Security and Defense Policy*, Brussels, Belgium: European Council, December 19, 2013, available from *www.consilium.europa.eu/uedocs/cms_data/docs/pressdata/en/ec/140214.pdf*.

8. Interestingly, the Baltic states, Poland, Britain, the Netherlands, Germany, the EDA, and NATO's Allied Command Transformation attended the meeting. There was specific discussion of developing cooperation with the Baltic nations. See *Nordic Defence Cooperation 2020*, December 4, 2013, available from *www.nordefco.org/Nordic-Defence-Cooperation-2020*.

9. A battle group is a multinational rapid-reaction force with separate internal command and logistics capabilities intended to provide a quick reaction capability.

10. HIP is a NATO designation for the Soviet-era Mi-8 transport helicopter.

11. See Ministry of Defence and Armed Forces of the Czech Republic, Multinational Aviation Training Centre Document Signed by Four Nations, February 25, 2013, available from *www.army.cz/en/ministry-of-defense/newsroom/news/multinational-aviation-training-centre-document-signed-by-four-nations-80184/*.

12. The Weimar Battle Group entered operational standby for the first time in the spring of 2013.

13. See UK Ministry of Defence, "UK-France Defence Co-operation Treaty Announced," November 2, 2010, available from *www.gov.uk/government/news/uk-france-defence-co-operation-treaty-announced--2*.

14. See Royal Navy, "Royal Navy Officers Join French Flagship's Gulf Deployment, March 3, 2014, available from *www.royalnavy.mod.uk/news-and-latest-activity/news/2014/march/03/140303-rn-french*.

15. See Ministry of Defence, "Double First for French and British Fast Jet Pilots," February 11, 2013, available from *www.gov.uk/government/news/double-first-for-french-and-british-fast-jet-pilots*.

16. See Pierre Tran, "UK, French Leaders Agree to Cooperate on Drone, Missile and More," *Defense News*, February 1, 2014, available from *www.defensenews.com/article/20140201/DEFREG01/302010025/UK-French-Leaders-Agree-Cooperate-Drone-Missile-More*.

17. For more information on the Belgium-Netherlands defense cooperation accords, see Advisory Council on International Affairs, European Defence Cooperation: Sovereignty and the Capacity to Act, January 2012, available from *www.gov.uk/government/uploads/system/uploads/attachment_data/file/224227/evidence-adviesraad-internationale-vaagstukken-european-defence-cooperation.pdf*.

18. See "Netherlands, Belgium and Luxembourg Enhance Defence Co-operation," *Dutch Daily News*, April 26, 2012, available from *www.dutchdailynews.com/netherlands-belgium-and-luxembourg-enhance-defence-co-operation/*.

19. For a concept paper regarding an expanded EPAF, see, "Regional Fighter Partnership: Options for Cooperation and Cost Sharing," Kalkar, Germany: Joint Air Power Competence Centre, March 2012, available from *www.japcc.org/portfolio/regional-fighter-partnership-options-for-cooperation-and-cost-sharing/*.

20. This chapter focuses on Europe's role in pooling and sharing. The United States and Canada can, of course, participate in such programs, but a primary objective should be to enhance European capabilities. Canada has played an outsized role in NATO's involvement in Afghanistan and takes its defense capabilities very seriously. For a Canadian perspective, see Mike Greenley, *Canadian Views on Smart Defence*, Ottawa, Ontario, Canada: Canadian Association of Defence and Security Industries, October 30, 2011, available from *www.ndia.org/Divisions/Divisions/International/Documents/_Greenley_CADSI_-_Greenley_-_Quadrilateral_-_Oct_2012[1].pdf*.

21. It must also be kept in mind that both NATO and the EU can only undertake those projects that member nations authorize and are willing to fund. In addition, there is at present an inherent limit on EDA activities, given the differing EU member views on the nature and extent of the EU security role.

22. This made it somewhat awkward for U.S. officials at NATO who were urging other nations to join the program while the United States was not doing so.

23. However, while there is regular interaction between NATO and the EU, much of it has been *pro forma*. Any truly significant enhancement of their relationship will most likely have to await resolution of underlying political issues such as the Turkey-Cyprus question.

24. This provision permits a few nations, rather than all EU nations, to cooperate. As noted earlier, to date, much of the EU pooling-and-sharing effort has consisted of analyses, communi-

qués, speeches, and initiatives. At the same time, through this process, the EU and the EDA have developed a wide-ranging and very ambitious menu of program priorities, with some more practical than others. Of course, the fact remains that all EDA efforts depemd wholly on having adequate resources made available and, as importantly, on national government decisions to provide funding and actual participation in specific activities. It remains to be seen how much of this agenda can and will go beyond studies and planning documents and be translated into active programs.

25. NATO is, in fact, developing a Connected Forces Initiative to build on connections established in Afghanistan.

26. Multinational storage of Grippen aircraft spare parts is already taking place among Sweden, the Czech Republic, and Hungary. Such pooling and sharing among European allies could also facilitate better burden sharing with the United States in meeting NATO capability and force generation requirements. There might be instances where the pooling and sharing should include the United States, such as intelligence, surveillance, and reconnaissance; special operations forces; and ballistic missile defense.

27. The principal NATO mechanisms for addressing industrial matters are the Conference of National Armaments Directors and the office of the NATO assistant secretary general for defense investment.

28. As noted previously, NORDEFCO is developing activities with Baltic nations. A forthcoming opportunity for cooperation will arrive with the acquisition by several European nations of the F-35 jet fighter. Norway and Britain have agreed to cooperate on their maintenance and use, and Norway will seek to extend that cooperation to the Netherlands.

29. For example, the very first signer of the memorandum of understanding for the SAC program was partner nation Sweden, and both Sweden and Finland have been active in various NATO military missions and activities.

30. It should be kept in mind that strong European military capabilities are in the U.S. national interest and have the potential to lessen the U.S. defense burden, especially given that U.S. defense capacities are becoming increasingly stretched. The United States does have very important allies and friends in Asia. Nonetheless, it remains the case that when push comes to shove, it is, above all, European nations that are most likely to join forces to assist the United States in military operations when U.S. security interests are at stake. In addition, NATO is the optimum mechanism for enhancing the military capacity of non-NATO nations that could contribute military forces to U.S.-led coalition military operations.

CHAPTER 13

JAPANESE HARD POWER: RISING TO THE CHALLENGE[1]

Toshi Yoshihara

KEY POINTS

- Japan's ambition to play a larger role on the world stage and address the security problems posed by a rising China have led Tokyo to undertake institutional, policy, and defense reforms.
- Japan's military reforms are intended to move its defense force from a posture of passive deterrence to one that is agile and forward leaning.
- Given Japan's budgetary restraints, however, it is unclear whether its resources can match its strategic ambitions.

No longer is Japan the political shrinking violet of the immediate post-war years. Historians will look back on the first decades of the 21st century as a turning point for Japanese strategy, both in East Asia and beyond. From major national security decisions—including the recent move to assume a limited right of collective self-defense—to a shift in military posture to counter a rising China, Japan is steadily loosening the constraints on its security policy. Japanese hard power, which includes Japan's first rate but constitutionally handicapped military, will correspondingly play a more prominent role in Tokyo's strategic calculus.

Understanding how Japanese policymakers will wield that hard power as an instrument of statecraft is thus crucial to Asian and global security. To explore how Japan's newfound assertiveness will shape Japanese hard power, this chapter will 1) assess recent developments in Japan's national security establishment and the deteriorating regional environment, 2) evaluate Japan's defense posture and military modernization efforts, and 3) identify the various financial and demographic constraints that could limit the material dimensions of Japanese strategy.

A "NORMAL" JAPAN AT LAST?

Prime Minister Shinzō Abe, who returned to power in late-2012 following the Liberal Democratic Party's landslide victory in the Diet's lower-house elections, has pushed aggressively to realize his ambitious agenda. Within a year of being elected, Abe instituted sweeping reforms to the national security apparatus. In December 2013, Japan announced the formation of a National Security Council (NSC) modeled after that of the United States. The council streamlines the prime minister's decisionmaking process while breaking down the various bureaucratic barriers that have impeded effective crisis management. Tokyo also enacted a controversial state secrets law that tightened the government's control over sensitive and classified information, enabling the NSC to centralize the handling of intelligence.

Concurrent with the NSC's creation, Tokyo issued three defense policy documents that furnish the roadmap for developing and sustaining Japanese hard power. The *National Security Strategy* (NSS), the first of its kind, sets forth "Japan's fundamental policies pertaining to national security."[2] The document is a welcome expression of Japan's long-term vision for securing the nation's regional and global security objectives. The fifth *National Defense Program Guidelines* (NDPG) establishes Japan's longer-term defense policy and force structure.[3] The *Medium Term Defense Program* (MTDP) is a programmatic statement of defense requirements and acquisition plans over a 5-year period.[4] For the first time in Japan's post-World War II history, Tokyo has produced policy documents that systematically align Japanese policy, strategy, and capabilities.

Notably, the NSS promotes the concept of "proactive contribution to peace" that commits Japan to an even more forward-leaning posture in world affairs. Describing the concept as a "fundamental principle of [Japan's] national security," the NSS argues that the security of Japan and of the wider international community have become indivisible:

> Japan cannot secure its own peace and security by itself, and the international community expects Japan to play a more proactive role for peace and stability in the world, in a way commensurate with its national capabilities.[5]

In other words, Japan advances global security by safeguarding its own neighborhood, while Japanese defense of the international order benefits Asian regional stability.

As such, the NDPG contends that Japan must:

> contribute even more proactively in securing peace, stability and prosperity of the international community while achieving its own security as well as peace and stability in the Asia—Pacific region.[6]

Indeed, Prime Minister Abe can look to Japanese contributions to international peace and security since the end of the Cold War as the basis for his foreign policy vision.

In a concrete manifestation of this proactive stance, the Abe administration relaxed Japan's arms exports ban, which had been in place for nearly 5 decades. Issued in April 2014, the new guidelines for transferring defense equipment intend to enhance technological cooperation with partners and friends, raising Japan's profile in regional and global arms markets. The move quickly bore fruit. A week after the new policy was announced, Australia and Japan agreed to a joint research project on marine hydrodynamics for constructing new submarines.

In July 2014, the newly established NSC approved Japan's research with Britain on the Meteor air-to-air missile and approved exporting a sensor component for the Patriot Advanced Capability-2 air defense system to the United States. A network of defense collaboration centered on developing hard power among like-minded nations could well emerge from these joint ventures. A proactive contribution to peace is thus as much about empowering other defenders of the status quo as it is about strengthening one's own capabilities.

In an even more consequential move, Abe partially lifted Japan's self-imposed ban on the right of collective self-defense, the hallmark of the nation's post-World War II foreign policy. For decades, successive Japanese governments strictly followed the constitutional interpretation that permitted Japan to exercise the right of individual self-defense, which forbids Japan's Self-Defense Forces (SDF) from aiding friendly or allied military units that have come under enemy assault.

This self-denial of a universal right, a right recognized under the United Nations (UN) charter, has long imposed a highly asymmetric and awkward arrangement on the U.S.-Japanese alliance. Washington would be obliged by treaty to defend Japan should it be attacked, while Tokyo could not reciprocate without violating its constitution. To Abe and his followers, such a legal constraint has become untenable in an increasingly dangerous security environment.

Among the scenarios used to advance Abe's initiative, two relating to the U.S.-Japan alliance stand out. Imagine that a Japanese warship were in the vicinity of an American naval unit under attack, and the warship took no action because of constitutional constraints. Imagine, too, that a Japanese destroyer equipped with the Aegis ballistic missile defense system were in a position to intercept a long-range missile headed for the United States, but the destroyer failed to do so, owing to Japan's ban on collective self-defense. To Abe and his lieutenants, if either of these crises occurred and Japan did nothing, then the alliance might not survive the subsequent political blowback in Washington. Thus, adopting the right to collective self-defense would signal Japan's determination to act alongside the U.S. military, sustaining the alliance's integrity while enhancing allied deterrence.

In July 2014, after intense negotiations with the New Komeito—the Japanese government's ambivalent junior coalition partner—Abe's cabinet approved the reinterpretation of the constitution, allowing Japan to nominally exercise its right of collective self-defense. Under the new understanding, use of force would be permitted "not only when an armed attack against Japan occurs but also when an armed attack against a foreign country that is in a close relationship

with Japan occurs."[7] However, in a compromise acknowledging the New Komeito's concerns, the Japanese government attached three key conditions necessary to invoke the right:

1. Only an attack or an impending attack that
"threatens Japan's survival and poses a clear danger
to fundamentally overthrow people's right to life, lib-
erty, and pursuit of happiness" would meet the constitutional standards for engaging in collective self-
defense.

2. Moreover, policymakers must determine that
"no other appropriate means" were available to counter the threat to Japan.

3. Even then, the SDF must limit its use of force to
"the minimum extent necessary" to repel or defeat the
threat.[8]

Abe's cabinet further acknowledged that "prior
approval of the Diet is in principle required upon issu-
ing orders" to the SDF for collective self-defense missions.[9] By no means, has Japan been unshackled from
its constitutional restraints or from its exclusively
defensive orientation.

The cabinet decision represents just the first step
in what will likely be a deliberate political process to
operationalize this broader constitutional interpretation. The Abe administration will need to submit a
legislative package to the Diet that would provide the
proper legal framework for the SDF to help assist or
defend allies and friends, should they come under attack. At least 10 existing laws would be reviewed, updated, and revised in this process. Opposition parties
will have another chance to litigate the issue.

In the meantime, changes in popular opinion or
other domestic political developments, such as local

election outcomes, could influence the momentum behind Abe's initiative. Public debate and legislative scrutiny—integral to Japan's open democratic system—will inevitably accompany this important shift in defense policy. Change will come incrementally through careful and transparent negotiations.

It is still unclear how the concept of limited collective self-defense will translate into operational practice for the U.S.-Japan alliance. Planned revisions to the U.S.-Japan defense guidelines, which spell out the allied division of labor, will reportedly incorporate an expanded defensive and logistical role for the SDF. Due for completion at the end of 2014, the guidelines called on the SDF to provide maintenance, supplies, and fuel to American military units heading into a combat zone—all rear-area activities that were previously prohibited.

In addition to improving allied cooperation, the cabinet decision could broaden the scope of the SDF's out-of-area operations. For example, the Abe administration has identified minesweeping as a potentially permissible action under UN Security Council authorization. Given Japan's dependence on energy from the Persian Gulf region, the mining of the Strait of Hormuz could constitute a clear threat to the nation's survival and well-being. This and other scenarios will likely be the subject of further debate when the government submits its legislative package to the Diet.

Japanese officials must strike a balance between adhering to the constraints of the cabinet decision and ensuring sufficient flexibility to account for the uncertainties of real-world military contingencies. Limited collective self-defense will open the door for Japanese hard power to play a more effective and meaningful role in maintaining regional and global security.

JAPAN'S NEIGHBORHOOD GETS ROUGHER

Mounting pressures close to home, including China's rise and North Korea's unpredictability, largely explain the quickening pace of Japan's normalization. China's assertiveness in the East China Sea over the past 5 years has been most troubling to Japan. In September 2010, China reacted vociferously after Japanese law enforcement arrested a Chinese fishing boat skipper who was filmed ramming Japanese Coast Guard vessels in the waters off the Senkaku Islands. Beijing used economic coercion, cutting off Japan's only supply of rare earth minerals critical to electronic manufactures.

After Tokyo nationalized the Senkakus in September 2012, Chinese maritime law enforcement flotillas began making the rounds in the disputed waters near the islands, and China has insisted that the regular patrols are routine. In response, Japanese Coast Guard vessels have been working overtime to monitor and trail every Chinese "intrusion," lest Tokyo concede Beijing's jurisdictional claims. Japan and China have been staring each other down in the East China Sea ever since.

Beyond the Senkakus dispute, Japan and China are locked in a budding naval rivalry. As China's rapidly modernizing navy extends its reach, it has become commonplace for Chinese naval flotillas to sail through Japanese-held narrow seas. Beginning in 2008 as sporadic forays into the Pacific, these expeditions now take place regularly year round. Moreover, the Chinese navy has steadily expanded the scope of its peacetime operations.

Notably, in July 2013, a surface action group steamed through the Sōya Strait (the first time Chinese units had conducted such a transit), circumnavigated Japan, and circled back to port by way of the international strait between Okinawa and Miyako Islands. Reflecting Tokyo's growing concerns about China's naval activism, Japan's annual defense white papers meticulously report the courses taken by Chinese naval task forces.

Chinese military aircraft, including fighter jets, have also ramped up flight operations over the East China Sea. In July, September, and October 2013, Y-8 airborne early warning aircraft and H-6 medium–range bombers conducted long range flight operations over the Pacific Ocean, passing between Miyako and Okinawa to reach the open sea.[10] Japan's Air Self-defense Force (ASDF) launched a record number of intercepts against Chinese aircraft in fiscal year 2013, surpassing the number of scrambles in fiscal year 2012 by more than 30 percent.[11]

In November 2013, Beijing unilaterally declared an Air Defense Identification Zone (ADIZ) over the East China Sea that requires all foreign aircraft entering the zone to submit flight plans to Chinese aviation authorities. The Chinese ADIZ pointedly overlaps with Japan's, extending to the Senkakus. Given that China is committed to making these increased naval and air activities the new status quo, frequent run-ins between Chinese and Japanese forces within the relatively confined spaces of East Asian seas will likely be the norm in the coming years.

Japanese policy documents routinely express Tokyo's misgivings about China's maritime rise. The NSS asserts:

China has taken actions that can be regarded as attempts to change the status quo by coercion based on their own assertions . . . in the maritime and aerial domains, including the East China Sea and the South China Sea.[12]

The NDPG further observes:

China has taken assertive actions with regard to issues of conflicts of interests in the maritime domain. . . . As for the seas and airspace around Japan, China has intruded into Japanese territorial waters, frequently violated Japan's airspace, and has engaged in dangerous activities that could cause unexpected situations.

The report singles out China's newly established ADIZ over the East China Sea as destabilizing, concluding, "As Japan has great concern about these Chinese activities it will need to pay utmost attention to them."[13]

Successive editions of the Japanese Defense Ministry's annual defense white papers have devoted more attention to China's maritime activism. In response to recent Chinese provocations at sea, the 2013 edition uses unusually blunt language to admonish Beijing:

Some of these activities of China involve its intrusion into Japan's territorial waters, its violation of Japan's airspace and even dangerous actions that could cause a contingency situation, which are extremely regrettable. China should accept and stick to the international norms.[14]

Since 2011, the defense ministry's internal think tank, the National Institute for Defense Studies, has published annual reports on China's security policy, offering a valuable regional perspective and a second opinion to the Pentagon's assessment of Chinese military power. Notably, the institute devoted the entire 2012 issue to Chinese maritime strategy and activities.[15]

In the meantime, North Korea refuses to fade into the background. In a series of provocations in 2010, North Korea sank the South Korean corvette *Cheonan*, revealed a new uranium enrichment facility, and shelled an island along the inter-Korean frontier. Pyongyang's ongoing development of its nuclear weapons and missile programs continues to pose a major security threat to Tokyo. North Korea has thus far conducted a nuclear test in 2006, 2009, and 2013. A fourth test will reportedly provide the reclusive regime sufficient data to design a nuclear warhead small enough to fit atop a ballistic missile.

Since the 1990s, North Korea has test-launched a series of ballistic missiles, with varying degrees of success. In December 2012, Pyongyang placed a satellite into orbit following a failed bid 8 months earlier. Widely seen as a disguise for a missile test, the successful space launch demonstrated North Korea's advances in long range rocketry and its potential ability to develop intercontinental ballistic missiles. After a nearly 5-year hiatus, the country resumed testing of its medium range Nodong ballistic missile, splashing two into the Sea of Japan in March 2014. With an estimated range of at least 1,000 kilometers, the Nodong can reach large parts of Japan. As the NDPG asserts:

> North Korea's nuclear and missile development, coupled with its provocative rhetoric and behavior, such as suggesting a missile attack on Japan, pose a serious and imminent threat to Japan's security.[16]

THE DYNAMIC JOINT DEFENSE FORCE

Chinese naval and air activities in and around the East China Sea and the employment of paramilitary maritime units near the Senkakus pose particularly taxing challenges for Japan. These peacetime tactics have enabled Beijing to apply constant pressure on Tokyo. China has thus far kept its frequent encounters with the SDF and Japanese Coast Guard at a low simmer, avoiding escalation, yet ensuring that the stand-off remains in play.

Short of capitulation, Japan has had no choice but to oblige in the cat-and-mouse game, lest it concede to China's jurisdictional claims or to its larger strategic aims in maritime Asia. Because Beijing has carefully calibrated its displays of force, Tokyo must respond judiciously to Chinese provocations. As China grows more powerful, this twilight phenomenon—featuring nervy close encounters falling well short of armed conflict—is likely to become a new "normal." Japan thus finds itself in a protracted contest of wills with no end in sight.

As the NSS observes:

> The Asia-Pacific region has become more prone to so-called 'gray zone' situations, situations that are neither pure peacetime nor contingencies over territorial sovereignty and interests. There is a risk that these 'gray zone' situations could further develop into grave situations.[17]

An incident at sea or a midair collision could trigger Sino-Japanese interactions that quickly spin out of control. In January 2013, a Chinese frigate locked its fire control radar on a Japanese destroyer, a threatening gesture that typically precedes weapons release.

Chinese fighters' dangerously close intercepts of Japanese surveillance aircraft in May and June 2014 lend credibility to fears that frequent military encounters could lead to accidents and even spiraling escalation. The NDPG further notes:

Amid the increasingly severe security environment surrounding Japan, the SDF, in addition to its regular activities, needs to respond to various situations, including 'gray zone' situations which require SDF commitment. The frequency of such situations and the duration of responses are both increasing.[18]

Tokyo clearly recognizes that China's aggression is not a passing phenomenon; rather, it will likely intensify in the coming years. As such, the NDPG asserts, "Japan will swiftly and seamlessly respond to situations including gray zone situations, and will establish the necessary posture to continuously address a protracted situation."[19]

To cope with the ambiguities and complex demands of gray zone contingencies, the NDPG pledges to:

build a Dynamic Joint Defense Force, which emphasizes both soft and hard aspects of readiness, sustainability, resiliency and connectivity, reinforced by advanced technology and capability for C3I, with a consideration to establish a wide range of infrastructure to support the SDF's operation.[20]

Such a force, according to the MTDP:

will provide an effective defense which enables the SDF to conduct a diverse range of activities based on joint operations seamlessly and dynamically, adapting to situations as they demand, while prioritizing particularly important functions and capabilities through optimal resource allocation.[21]

Despite the impenetrable jargon typical of defense reports, these stated objectives provide a roadmap to the SDF's modernization programs and future force structure.

The Dynamic Joint Defense Force concept is the product of a steady evolution in Japanese strategic thought. Notably, the 2010 NDPG formally jettisoned the Basic Defense Force Concept, a Cold War legacy premised on strong, yet relatively immobile, defenses designed to repel assault and predicated on a largely passive deterrence posture. Instead, according to the 2010 NDPG, a dynamic defense force would take the place of static defense, and agility would be the watchword of the new force. Such forces could swiftly deploy to remote islands for a variety of contingencies, meeting challenges as they arose. To develop a dynamic defense force, the SDF would concurrently rejuvenate aerial, surface, and underwater surveillance operations.

The dynamic joint defense force thus carries forward many of the key tenets developed in 2010. In addition to mobility and readiness, the 2013 NDPG emphasizes the close coordination among naval, air, and ground forces. The inherently amphibious character of the Japanese-held islands in the East China Sea demands such integration of capabilities. At the same time, the 2013 NDPG calls on the SDF to establish an effective intelligence, surveillance, and reconnaissance (ISR) architecture that would blanket the East China Sea with a variety of sensors to better monitor China's naval and air activities. All three services would benefit enormously from such an enhanced ability to keep track of Chinese forces. To meet the ambitions of the dynamic joint defense force concept, the SDF has embarked on a series of military modernization programs.

FORCE MODERNIZATION TRENDS

Japan boasts one of the most modern and professional militaries in the world. During the Cold War, the SDF complemented – and filled the gaps of – the U.S. military presence in the Western Pacific. Japan's armed forces shielded the home islands while the major forward bases along the Japanese archipelago allowed the United States to project power across Asia and beyond. The Japanese Maritime Self-defense Force's (MSDF) surface, undersea, and air units bottled up Soviet naval forces in the Sea of Japan. The maritime service also kept open the sea lanes and secured the maritime approaches to Japan, which were critical to the nation's economic well-being.

The ASDF's modern fighters ensured that Japan could defend the airspace over and near the country. The nation's Ground Self-defense Force (GSDF) bristled with tanks and artillery to defend against a full-scale Soviet invasion of the homeland, particularly against Hokkaido Island. The SDF was – and remains – largely a defensive force designed to maintain the nation's territorial integrity, possessing limited offensive power projection capabilities. Japan's current force structure and posture are thus legacies of this superpower rivalry.

As a quintessential maritime nation, it is not surprising that Japan counts the MSDF as its leading service. Over the past decade, Japanese naval power has evolved in both quantitative and qualitative terms. In

2010, Tokyo announced its plan to increase its world class submarine fleet from 16 to 22 boats, representing a nearly 40 percent jump in size. The decision was all the more remarkable because the number of boats had stayed fixed at 16 since 1976.

Leading this growth is the cutting-edge Sōryū-class diesel-electric submarine. The largest of its kind in the world, the Sōryū is superior to its predecessor by virtually every index of performance. It is the first Japanese boat fitted with air independent propulsion, a fuel-cell technology that permits submarines to operate underwater for extended periods while quieting their noise signature. In short, the MSDF leads the region in conventional submarine warfare, constituting the benchmark against which other Asian navies will be compared over the next decade.

Notably, Japan has been able to invest in its undersea prowess without imposing undue burdens on its fiscal position. The MSDF has traditionally decommissioned its submarines unusually early, introducing more advanced boats to replace older ones that could have stayed in active service for at least another decade. To support the current buildup, the maritime service began keeping its existing boats at sea longer, allowing for a steady growth in fleet size without substantially increasing acquisition costs. Japan will likely meet its 22-boat target before the end of the decade.

The MSDF's surface fleet, comprised of nearly 50 major surface combatants, has also undergone a makeover. In 2009, the maritime service commissioned the first of two Hyūga-class helicopter carriers with a full load displacement of 19,000 tons. Capable of embarking as many as 11 helicopters, the carrier is a powerful antisubmarine warfare (ASW) platform.

In 2013, Japan launched the first of two Izumo-class ASW helicopter carriers that displace 27,000 tons at full load and carry up to 14 helicopters. Measuring nearly 250 meters in length, the Izumo will enter service in 2015 as the largest warship the Japanese have built since World War II. It promises to boost substantially Japan's ability to conduct and sustain ASW operations alongside the Hyūga-class carriers.

Additionally, two more Aegis-equipped surface combatants will join the four Kongō-class and two Atago-class guided missile destroyers to enhance Japan's missile defense capabilities at sea. In 2012, the first of four Akizuki-class guided missile destroyers was commissioned to provide anti-air, anti-surface, and anti-submarine cover for the helicopter carriers and Aegis-equipped destroyers. For the MSDF's air fleet, the P-1s—the next generation maritime patrol aircraft—will eventually replace the aging P-3Cs as Japan's main shored-based, fixed-wing ASW unit.

The ASDF fields a mix of fourth— and third—generation fighters, including nearly 200 F-15s, 90 F-2s (a variant of the American F-16), and 60 F-4s. A modest number of KC-767 aerial refueling tankers, E-767 Airborne Warning and Control System aircraft, and E-2C airborne early warning aircraft provide support to Japan's air superiority and multirole combat aircraft. A fleet of C-130 and C-1 transports furnishes limited strategic lift to Japanese forces.

The most prominent and expensive modernization program for the air service is that of the fifth-generation F-35 fighters. Because of the prohibitive per unit cost of the aircraft, which has risen further with Japan's participation in the local production of the fighters' parts, the ASDF currently plans to acquire only 42 F-35s. (See Table 13-1.)

Japan is also developing its own stealth fighter, the Advanced Technology Demonstrator-X, to replace the F-2s. The C-2 transport, the successor to the C-1, promises to improve substantially the range and capacity of the ASDF's lift. Japan's air service will acquire new airborne early warning aircraft, aerial refueling tankers, and transports to augment the ASDF's ability to patrol the airspace around the Japanese islands. Japan will also invest in UAVs—a joint asset available to the three services—to enhance its ISR capabilities. The leading contender to enter service with the ASDF is the high altitude, long endurance Global Hawk.

The GSDF is undergoing the most dramatic restructuring and reorganization of recent years. Reflecting Tokyo's judgment that the risk of a homeland invasion is negligible, about 700 main battle tanks and 600 artillery pieces will be reduced to 300 and 300, respectively, over a 10-year period. Tank and artillery units will also be removed from Honshu Island and consolidated on Hokkaido and Kyushu Islands. (See Table 13-1.)

	1995 NDPG	2013 NDPG	Future
SDF			
Active Duty Personnel	145,000	151,000*	151,000
GSDF			
Tanks	900*	700*	300*
Artillery	900*	600*	300*
MSDF			
Destroyers	50*	47	54
(Aegis-equipped destroyers)	--	6	8
Combat aircraft	170*	170*	170*
Submarines	16	16	22
ASDF			
Combat aircraft	400*	340*	360*

Source: Japanese Ministry of Defense, "Defense of Japan 2014," Figures II-4-3-2 and II-4-3-3, available from *www.mod.go.jp/e/publ/w_paper/pdf/2014/DOJ2014_Figure_1st_0730.pdf*.

Notes: An asterisk denotes approximate figures. The "Figure" column derives from the 2013 NDPG's discussion of a future defense posture that will probably be achieved within a 10-year period.

Table 13-1. NDPG Comparison of Personnel and Equipment.

To enhance responsiveness and mobility, the GSDF will form two rapid-deployment divisions and two rapid-deployment brigades. Most notably, the ground service will create a new marine brigade capable of conducting amphibious operations to retake remote islands seized by enemy forces. Japan will procure the AAV-7 amphibious assault vehicles and V-22 tiltrotor aircraft that would provide Japanese marines with organic lift capability to project forces ashore.

It is worth noting that these modernization efforts will likely strengthen the SDF's capacity to project only limited power in the coming years. Notwithstanding breathless commentary surrounding the unveiling of the Izumo-class helicopter carrier, the MSDF is many steps and years away from acquiring a fixed-wing carrier strike force. Long range bombers or intercontinental ballistic missiles are conspicuously missing from the ASDF's inventory, and the GSDF can only conduct limited expeditionary operations for territorial defense.

The SDF is still very much the shield that counts on the American spear to fulfill the full range of missions in Japan's defense. This is consistent with Tokyo's current constitutional interpretation prohibiting the possession of weaponry capable of prosecuting offensive operations. Any attempt to depart from this defensive orientation will not escape Japan's democratic processes, involving painstaking negotiations and debates. Fears of creeping Japanese militarism are thus unwarranted.

DEFENDING THE SOUTHWEST ISLANDS

As successive policy documents make clear, Tokyo will strengthen its defense posture along the Ryūkyū Islands in the southwest, the geographic epicenter of the Sino-Japanese rivalry. By beefing up defenses along the Ryūkyūs, Japan might be able to exploit a permanent geographic advantage. The island chain gives the SDF the option of closing off Chinese access to the high seas—much as Japan's Home Islands formed a physical barrier that kept the Soviet Navy bottled up in the Sea of Japan—and provide a form of strategic leverage. Indeed, given Beijing's deeply embedded fears of being denied access to the global commons, a powerful blocking force along the Southwest Islands could bolster Japan's deterrence posture.[22]

The MTDP directs the GSDF to establish a new coastal reconnaissance unit on Yonaguni, the westernmost island of the Ryūkyū archipelago, strategically located about 70 miles east of Taiwan and about 100 miles southwest of the Senkakus.[23] A garrison on Yonaguni would extend Japan's situational awareness to its largely undefended and potentially vulnerable southern flank. In a cross-Taiwan Strait war, for example, Chinese forces would likely transit the seas and airspace near Yonaguni to attack Taiwan's less-defended east coast. The ASDF will redeploy an airborne early warning squadron and a fighter squadron to Naha Airbase in Okinawa, reinforcing the squadrons already there. The MSDF will refit its three Ōsumi-class tank landing ships to accommodate the planned purchases of the MV-22 tiltrotor aircraft and the AAV-7 amphibious assault vehicle. In June 2013, in an early sign of Japanese intentions, a U.S. Marine Corps MV-22 landed on the Hyūga helicopter carrier during an allied exercise.

Japan is also applying lessons learned from its Cold War experiences. In anticipation of a massive Soviet amphibious assault on Hokkaido Island, the Japanese developed an anti-invasion strategy that employed shore-based missile units to strike approaching enemy transports. Tokyo is now replicating this asymmetric tactic in the south. The 2013 NDPG calls on the GSDF to "maintain surface-to-ship guided missile units in order to prevent invasion of Japan's remote islands while [invading forces are] still at sea."[24]

Two years earlier, the GSDF deployed several units armed with Type 88 anti-ship cruise missiles (ASCMs) to Amami Ōshima, which is near the northern end of the Ryūkyūs. In November 2013, the GSDF put ashore Type 88 missiles on Miyako Island as a part of a larger military exercise. These unprecedented shows of force were no doubt directed at Beijing, as Chinese naval flotillas frequently transit the strait between Miyako and Okinawa Islands. The message was not lost on the Chinese.

The GSDF's truck-launched Type 88 ASCM makes for an ideal weapon on the Southwest Islands. With a range of 110 miles, Type 88s can strike warships at sea from sites far inland. Well-placed ASCM batteries could cover all Ryūkyū narrow seas, while converting the eastern edge of the East China Sea into a no-go area for Chinese surface forces. The GSDF has begun acquiring the Type 12 ASCM, the successor to the Type 88.[25] Boasting greater reach, precision, and survivability, these new missile units promise to render transiting straits or nearby waters even more perilous for Chinese mariners.

As the Dynamic Joint Defense Force concept illustrates, an effective defense of the Ryūkyūs would require unprecedented coordination among the three services. The GSDF's amphibious forces and shore-based anti-ship missile units would rely on the lift capabilities of the ASDF's air transports and the MSDF's vessels to reach rapidly islands stretching over 1,000 kilometers between Kyushu Island and Taiwan. The coastal reconnaissance garrison on Yonaguni would provide early warning to air and naval units. The Type 88 and Type 12 ASCM launchers would require the cueing and targeting data from the MSDF's airborne early warning aircraft to conduct over-the-horizon strikes against enemy surface forces. When they enter service, ASDF UAVs would enhance the situational awareness of all units operating around the Southwest Islands. Above all, Japanese warships, submarines, and fighters must ensure sea control and air superiority, without which amphibious operations and island defense would founder. Mutual support among the three services in a complex operational environment is thus essential to success.

A CONVENTIONAL COUNTERSTRIKE OPTION?

If the military balance continues to tilt in Beijing's favor, Tokyo could feel compelled to deter by punishment, which could entail inflicting unacceptable levels of pain on China should the People's Liberation Army (PLA) ever attack Japan and Japanese forces. To retaliate directly against China with such force, Japan would have to develop offensive strike capabilities designed to hold at risk a range of assets, especially those on the mainland that Beijing highly values. In theory, Tokyo's ability to impose prohibitive costs on China would deter the Chinese military from acting in the first place. Dating back to the 1950s, Japanese debates about the constitutionality of attacking enemy territory suggest that a decision to pursue deterrence by punishment is not far-fetched.

While an offensive posture would no doubt stoke political controversy, serious debates about acquiring land attack cruise missiles have surfaced in Japan from time to time since at least 2005.[26] The discourse has centered primarily on the legalities of Tokyo's hypothetical decision to attack North Korean missile bases in the event of a crisis. But it can be assumed that Japan would not limit the missile's use to Pyongyang if Japan ever acquired such a weapon system.

In 2009, the subcommittee of the defense policy-making council of the Liberal Democratic Party (LDP) submitted a proposal endorsing the acquisition of offensive missiles. The committee called on Japan to "maintain the capability to attack enemy missile sites" and recommended developing cruise and ballistic missiles and the space-based systems to support missile operations.[27] The LDP's electoral defeat in September 2009 ended further discussions on this issue.

Nevertheless, the report represented a significant milestone in post-war Japanese thinking about defense and helped legitimize the notion of going on the offense. Prime Minister Abe's electoral victory has resurrected the debate. In reference to the North Korean missile threat, the latest NDPG and the MTDP obliquely hint at revisiting a counterstrike capability. The NDPG states:

> Based on appropriate role and mission sharing between Japan and the U.S., in order to strengthen the deterrent of the Japan-U.S. alliance as a whole through enhancement of Japan's own deterrent and response capability, Japan will study a potential form of response capability to address the means of ballistic missile launches and related facilities, and take means as necessary.[28]

In other words, all options are back on the table. What would a conventional missile option look like? Tokyo would almost certainly limit itself to counterforce strikes aimed exclusively at enemy military units. This would require Japan to plan for counter-offensive operations against Chinese military forces, including those deployed on the mainland.

Equipping Japanese forces with conventional long range precision-strike weapons, such as the venerable

Tomahawk land attack cruise missile, would not only be relatively affordable but also technically feasible. In particular, Japanese destroyers, submarines, and aircraft armed with Tomahawks or their equivalents could strike large fixed targets, such as the over-the-horizon radars, essential for conducting Chinese anti-access operations. As Chinese dependence on land-based sensors to effectively employ its theater-strike systems increases, Japan may find the strategic dividends of a counterstrike capability operationally attractive and, thus, politically persuasive.

DEFENSE BUDGET WOES

While Tokyo's modernization plans are well tailored to address China's growing challenge, Japan may have trouble sustaining or expanding them to keep up with the Chinese military. On paper, Japan's annual defense budget, at nearly $48.6 billion in 2013, is impressive.[29] Indeed, Japan is ranked fifth in the world in military expenditures, following the United States, China, Russia, and France. But such a high figure paints a superficial picture at best. For decades, Japan capped its defense budget at 1 percent of gross domestic product (GDP)—far below figures expected of great powers—as an expression of its pacifist orientation.

Although Tokyo is not legally committed to such fiscal constraint, longstanding practice has formed a powerful normative prohibition against shattering this ceiling. Consequently, the fixed defense budget has plateaued alongside anemic economic growth since the early-1990s. Moreover, Japanese government debt is nearly 250 percent of GDP, and soaring social security expenditures owing to Japan's rapidly aging society have intensified competition over ever scarcer financial resources. Such fiscal burdens could prove crippling in the years ahead, draining the political will to spend substantially more on defense.

The past decade's budgetary trends reflect Japan's monetary predicament. The defense budget suffered cuts for 11 consecutive years, dropping from ¥4.94 trillion in 2002 to ¥4.64 trillion in 2012.[30] In 2013, Prime Minister Abe's government announced a very modest 0.8 percent budgetary increase over the previous year, reversing the steady decline. The cabinet then approved a 2.8 percent boost to its defense budget for fiscal year 2014, the largest year-on-year increase since the mid-1990s.

In light of the deteriorating security environment, the decision to reverse the steady decline was long overdue. While the spending hikes are welcomed news, they are unlikely to provide sufficient relief. Military modernization programs will compete with other priorities. For example, compensation for government pay cuts following the March 2011 earthquake, tsunami, and nuclear disasters could largely nullify the growth in outlays. It thus remains unclear how much more capability these modest increases will buy.

Japan's budgetary woes are even more alarming in comparative terms. China has dramatically surpassed Japan in defense spending over the past decade. The Stockholm International Peace Research Institute estimates that the Chinese defense budget, measured in constant 2011 dollars, grew from $52.8 billion in 2002 to $159.5 billion in 2012. Japan, by contrast, virtually stood still, with its budget declining slightly from $60.7 billion to $59.5 billion over the same period.[31]

The Japanese Ministry of Defense reckons that China's defense spending grew by 350 percent from 2003 to 2012, while Japan's budget shrank slightly during that decade.[32] The International Institute for Strategic Studies paints a similarly stark picture. In 1990, Japan spent, in nominal terms, nearly $29 billion on defense compared to China's $6 billion. By 2013, Chinese expenditures soared to $112 billion, more than doubling Japan's $51 billion defense budget.[33] (See Figure 13-1.) Such an extraordinary reversal in fortunes between two rival regional powers is rare by historical standards. Ominously, Japan's persistently low economic growth rates will likely permit China to further widen the spending gap.

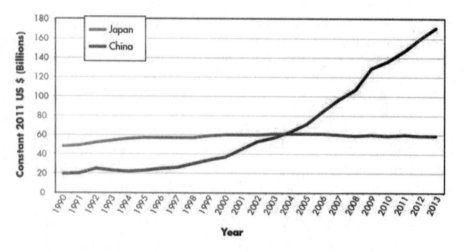

Source: Stockholm International Peace Research Institute (SIPRI), *Military Expenditure Database*, Stockholm, Sweden: SIPRI, available from *www.sipri.org/research/armaments/milex/milex_database/milex_database*.

Figure 13-1. Japanese versus Chinese Defense Expenditures.

STRATEGY—RESOURCE MISMATCH?

Beyond budgetary constraints, Japan's long-standing ambitions to fulfill wider international responsibilities befitting a major power—captured by Abe's concept of "proactive contribution to peace"—could spread the SDF too thin. Since Japan's dispatch of minesweepers to the Persian Gulf after the First Gulf War in 1991, successive Japanese administrations have deployed ground, air, and naval forces far beyond Japan's own neighborhood to conduct "international peace cooperation operations."

The 2013 MTDP defines such operations as:

activities cooperatively carried out by the international society to improve the international security environment such as UN Peace Keeping Operations, Humanitarian Assistance/Disaster Relief (HA/DR), and others in the fields of non-traditional security.[34]

Since Japan's first peacekeeping mission in Cambodia in 1992, Japan has sent peacekeepers around the world, including to the Golan Heights in the Levant, to South Sudan, to East Timor, and to Haiti. Japanese forces distinguished themselves in rendering assistance to stricken nations following the 2004 Indian Ocean tsunami and to the Philippines in the aftermath of the 2013 Haiyan typhoon.

In a post-September 11, 2001, show of solidarity with the United States, Tokyo committed MSDF vessels to the coalition naval contingent supporting combat operations in Afghanistan. MSDF tankers resupplied coalition warships, and Aegis destroyers guarded against air and surface threats in the Arabian Sea. MSDF vessels supplied fuel oil and water to customers from about a dozen countries—including the United States, Pakistan, France, Britain, and Germany—until the mission lapsed in January 2010.

Over the past decade, Japan has participated in various global efforts to secure peace. The country was a founding participant in the Proliferation Security Initiative in 2003 and has remained one of the initiative's foremost proponents. Moreover, a modest-sized GSDF contingent deployed to Iraq in January 2004 for noncombat duty. Tokyo joined the fight against Indian Ocean piracy in July 2009, committing to an open-ended, out-of-area deployment. Finally, Japanese mariners continue to ply the anarchic Gulf of Aden and Arabian Sea alongside a multinational contingent of naval forces. Tokyo subsequently established a military base at Djibouti to support forward deployed MSDF units, Japan's first overseas base since World War II.

While these praiseworthy activities have set a powerful precedent for fulfilling Prime Minister Abe's wider agenda, competing imperatives will likely force Japan to prioritize narrower national interests above global security. As the security environment deteriorates closer to home, Japan's willingness to spend political and military capital on extraregional missions will diminish commensurately.

Moreover, the Japanese government will place greater weight on managing direct threats to sovereignty and material prosperity than on meeting abstract, diffuse challenges in regions where Japan remains a marginal player. As an economically dynamic, militarily strong China eyes the Senkakus and the wider East China Sea, Japan's SDF, which is already inferior in numbers to the PLA, is losing its edge in its main East Asian theater, even as threats to Japanese interests in other parts of the world remain remote, ill-defined, and of indefinite duration. Whether the Japanese government can allocate resources deftly enough to balance traditional against nontraditional military functions remains uncertain. If Tokyo fails to prioritize, then it is entirely possible that Japanese political ends will outstrip ways and means.

JAPAN'S LOOMING DEMOGRAPHIC CRISIS

Over the long term, Tokyo will confront a structural and virtually irreversible challenge to its hard power. Japan's rapidly aging society is pushing the nation toward an unprecedented demographic crisis that could have dire implications for its defense posture in the coming decades. Owing to low fertility, high life expectancy, and trifling immigration, Japan will be significantly older and smaller in 2030 than it is today. The population will likely decline from 128 million in 2010 to 116 million 20 years hence, averaging a loss of more than 660,000 Japanese citizens per year.

During this same period, Japan's working-age population (ages 15 to 64) will shrink by 17 percent, from 81 million to 67 million. The median age of the population will rise from 45 to 50, and about a third of the population will be over 65 years old by 2030. Some forecasts estimate that Japan's population may shrink to 90 million by mid-century, representing an astounding 30 percent decrease from its peak years in the late-2000s.[35]

Population decline inevitably reduces the pool of manpower available for military service. The figures are sobering. The male population eligible to join the SDF (ages 18 to 26) peaked at 9 million in 1994. In just over 15 years, this age group recorded 30 percent drop, plummeting to around 6 million.

By 2030, SDF-eligible males will fall to less than 5 million. By contrast, the United States will likely experience a 16 percent increase for the same cohort between 2010 and 2030. The cost of fielding troops for combat will rise as manpower availability dwindles. In the coming years, maintaining satisfactory levels of recruitment and retention will likely tax Tokyo's resources. Indeed, the 2013 NDPG specifically cites the declining birthrate as a factor in pressurizing the recruiting environment.[36]

Recent defense policy documents have held out hope that technology will potentially lessen the effect of personnel shortages. But most military operations—ranging from high-end conventional wars to post-conflict reconstruction—soak up manpower. Gee whiz technologies, such as unmanned systems, only go so far. Warfighters in the field and support crews in the rear must still do much of the heavy lifting. Japan's response to the March 2011 tsunami was the starkest reminder of this reality: Tokyo called up more than 100,000 military personnel—about 40 percent of the active duty force—for relief operations, the largest deployment of troops in Japan's postwar history. In short, boots on the ground still count as much in peacetime as they do in war.

Unless Japan is prepared for a major military buildup, which appears politically doubtful and fiscally unsustainable, the country's shrinking pool of manpower will weigh heavily on Japanese decision-makers. It remains to be seen whether such socio-economic pressures will increase temptations to turn inward, even as Japan's external security environment grows more contentious.

WILL JAPAN RISE TO THE CHALLENGE?

China's rapid ascent has spurred Japanese policy-makers to reallocate considerable material resources and to expend intellectual energy on hard power during an era of fiscal austerity. Chinese behavior in recent years suggests that the stakes now involve nothing less than Japan's and China's future places in maritime Asia. At the very least, Tokyo's choices have narrowed: it can either accommodate Beijing in the near future, or it can act now to preserve the freedom of action it has enjoyed for decades. Not since the 1969 Richard Nixon Doctrine—a presidential call to America's Asian allies to protect themselves against external threats, even as the United States retrenched—has Japan confronted such strategic danger and stark options.

Only sound strategy will help Tokyo navigate the uncertainties of living in an unstable security environment. The extent to which Japan can shape its hard power to serve an effective strategy will depend on meaningful progress along multiple fronts. Tokyo must pivot even more decisively away from its northward orientation toward Russia—an anachronistic Cold War legacy—and toward its southern flank along the Ryūkyūs. Japan must stubbornly hold the line there, maintaining high levels of alertness, even while keeping its cool in the face of persistent Chinese probes and provocations. To do so, the SDF must develop unprecedented levels of cooperation and trust among its services to secure an extended front far from the Home Islands. Above all, Tokyo must sustain the political will and invest in the capabilities necessary to take up the Chinese challenge. Only thus can Japan hope to stay in a competition that promises to be a long one.

ENDNOTES — CHAPTER 13

1. This chapter was originally published as an essay on August 25, 2014.

2. Prime Minister's Cabinet and National Security Council, *National Security Strategy*, London, UK, December 17, 2013, p. 1, available from *www.cas.go.jp/jp/siryou/131217anzenhoshou/nss-e.pdf*.

3. The NDPG was previously published in 1976, 1995, 2004, and 2010.

4. Prime Minister's Cabinet and National Security Council, *Medium Term Defense Program, FY2014-FY2018*, London, UK, December 17, 2013, available from *www.mod.go.jp/j/approach/agenda/guideline/2014/pdf/Defense_Program.pdf*.

5. Prime Minister's Cabinet and National Security Council, *National Security Strategy*, pp. 3-4.

6. Prime Minister's Cabinet and National Security Council, *National Defense Program Guidelines for FY 2014 and Beyond*, London, UK, December 17, 2013, p. 5, available from *www.mod.go.jp/j/approach/agenda/guideline/2014/pdf/20131217_e2.pdf*.

7. Ministry of Foreign Affairs of Japan, *Cabinet Decision on Development of Seamless Security Legislation to Ensure Japan's Survival and Protect Its People*, Tokyo, Japan, July 1, 2014, p. 8, available from *www.mofa.go.jp/fp/nsp/page23e_000273.html*.

8. *Ibid.*

9. *Ibid.*

10. For detailed flight paths of Chinese aircraft over the East China Sea and through the Ryūkyū Islands, see Japanese Ministry of Defense, "China's Activities Surrounding Japan's Airspace," Tokyo, Japan, available from *www.mod.go.jp/e/d_act/ryouku/index.html*.

11. Kosuke Takahashi and James Hardy, "Japan Sees Big Rise in Scrambles against Chinese Aircraft," *HIS Jane's*, Vol. 360, April 9, 2014.

12. Prime Minister's Cabinet and National Security Council, *National Security Strategy*, p. 12.

13. For block quote and this quote, see Prime Minister's Cabinet and National Security Council, *National Defense Program Guidelines*, pp. 3-4.

14. Japanese Ministry of Defense, *Defense of Japan 2013*, July 2013, p. 39, available from *www.mod.go.jp/e/publ/w_paper/2013.html*.

15. *NIDS China Security Report 2011*, Tokyo, Japan: National Institute for Defense Studies, February 2012, p. 4, available from *www.nids.go.jp/english/publication/chinareport/pdf/china_report_EN_web_2011_A01.pdf*.

16. Prime Minister's Cabinet and National Security Council, *National Defense Program Guidelines*, p. 3.

17. Prime Minister's Cabinet and National Security Council, *National Security Strategy*, p. 11.

18. Prime Minister's Cabinet and National Security Council, *National Defense Program Guidelines*, p. 7.

19. *Ibid*. p. 13.

20. *Ibid*. p. 8.

21. Prime Minister's Cabinet and National Security Council, *Medium Term Defense Program*, p. 1.

22. See Michael Pillsbury, "The Sixteen Fears: China's Strategic Psychology," *Survival: Global Politics and Strategy*, Vol. 54, No. 5, October-November 2012, pp. 152-154.

23. Prime Minister's Cabinet and National Security Council, *Medium Term Defense Program, FY2014-FY2018*, p. 7.

24. Prime Minister's Cabinet and National Security Council, *National Defense Program Guidelines*, p. 22.

25. Japanese Ministry of Defense, *Defense Programs and Budget of Japan: Overview of FY 2014 Budget*, March 20, 2014, p. 8, available from *www.mod.go.jp/e/d_budget/pdf/251009.pdf*.

26. See, for example, Sugio Takahashi, "Strike Capability against Enemy Territory under Exclusive Defense Policy: One Option for Coping with Enemy Missile Threats," *NIDS Security Studies*, Vol. 8, No. 1, October 2005, pp. 105-121; and Hideaki Kaneda, "Is It Possible for the SDF to Attack Enemy Missile Bases?" *Sekai No Kansen*, February 2007.

27. National Defense Division of the Liberal Democratic Party, Subcommittee of the Defense Policymaking Council, *A Proposal Regarding the New National Defense Program Guidelines*, London, UK, June 6, 2009, pp. 11-12.

28. Prime Minister's Cabinet and National Security Council, *National Defense Program Guidelines*, p. 20.

29. The defense budget is given in 2013 dollars. For more information, see *Military Expenditure Database*, Stockholm, Sweden: SIPRI, available from *www.sipri.org/research/armaments/milex/milex_database*.

30. Japanese Ministry of Defense, *Defense of Japan 2013*, p. 119.

31. Data drawn from SIPRI *Military Expenditure Database 2014*, available from *milexdata.sipri.org/files/?file=SIPRI+military+expenditure+database+1988-2013.xlsx*.

32. Japanese Ministry of Defense, *Defense of Japan 2013*, p. 122.

33. I thank Bradford Lee for this insight. See *The Military Balance 1991*, London, UK: International Institute for Strategic Studies, pp. 150, 165; and *The Military Balance 2014*, London, UK: International Institute for Strategic Studies, pp. 230, 250.

34. Prime Minister's Cabinet and National Security Council, *Medium Term Defense Program*, pp. 1-2.

35. *World Population Prospects: The 2008 Revision 1*, New York: UN Population Division, 2008, pp. 292-293.

36. Prime Minister's Cabinet and National Security Council, *National Defense Program Guidelines*, p. 26.

CHAPTER 14

NORTH ATLANTIC TREATY ORGANIZATION AIR POWER: A SELF-RELIANT EUROPE?[1]

Craig Franklin

KEY POINTS

- With a decreasing U.S. Air Force presence in Europe and increasing pressure to address security concerns in Asia and the Middle East, non-U.S. North Atlantic Treaty Organization (NATO) air forces must shoulder more of the burden in Europe and its periphery.
- The strength of non-U.S. NATO air forces lies in their personnel, tactical fighter strength, and basing infrastructures.
- These air forces have plans to address key shortfalls in intelligence, surveillance, and reconnaissance; transport; air refueling; and stealth aircraft, but successful implementation will depend on Smart Defence initiatives and stable budgets.
- More broadly, two issues continue to hamstring NATO planning and execution: the fact that the alliance lacks a common understanding of the threats it faces, and the trend of NATO members placing caveats on the types of missions they will fly.

Although it is unlikely that NATO would ever participate in a conflict without significant airpower contributions from the United States, cuts to American capabilities necessarily lead one to wonder what America's NATO partners can bring to the table. Certainly, the U.S. Air Force presence in Europe is nowhere near what it once was—or even what it was in the 1990s.

In the 1990s, the U.S. Air Force in Europe had 25 main operating bases, 34 aircraft squadrons, and approximately 72,000 personnel. Today, there are just five main operating bases, eight aircraft squadrons, and approximately 25,000 personnel. Logically, this should mean that an ever-increasing part of any NATO air effort would be non-U.S. NATO's Smart Defence concept encourages NATO nations to shoulder a greater share of defense and to not just rely on U.S. capabilities. NATO describes the origins of Smart Defence as follows:

> From 2008 the world economy has been facing its worst period since the end of the Second World War. Governments are applying budgetary restrictions to tackle this serious recession, which is having a considerable effect on defense spending.
>
> Furthermore, in the course of this crisis, the Alliance's security environment has been changing, and has become more diverse and unpredictable. The crisis in Libya is a recent example, underlining the unforeseeable nature of conflicts, but also showing the need for modern systems and facilities, and for less reliance on the United States for costly advanced capabilities.
>
> In these crisis times, rebalancing defense spending between the European nations and the United States is more than ever a necessity. The other Allies must reduce the gap with the United States by equipping themselves with capabilities that are deemed to be critical, deployable, and sustainable, and must dem-

onstrate political determination to achieve that goal. There must be equitable sharing of the defense burden. Smart Defense is NATO's response to this.[2]

This chapter provides an overview of non-U.S. NATO air capabilities. It assesses the current state of non-U.S. NATO command and control (C2), airmen, aircraft, munitions, basing, air and missile defense and readiness. It concludes with 10 challenges facing America's NATO partners in these fields and outlines how NATO is or should be addressing each.

C2

NATO's Allied Air Command at Ramstein Air Base, Germany, is the singular NATO command for organizing air operations and is led by a four-star U.S. Air Force officer who is a dual-hatted U.S. and NATO commander. His Allied Air Command staff comes from a variety of NATO nations and currently includes a French three-star general as vice commander; a German two-star general as chief of staff; and Turkish, American, and British one-star generals as, respectively, deputy chiefs for plans, operations, and support.

Presently, the command has established nine focus areas: NATO charter Article 5 operations, NATO-integrated air and missile defense, NATO air policing, ballistic missile defense (BMD), support to ongoing NATO operations, Air Command (AIRCOM) joint force air-component readiness, partnerships with non-NATO member states' air forces, air and space advocacy, and air-capability development.[3]

Using lessons it has learned from NATO operations in the Balkans in the 1990s and Libya in 2011,

NATO has streamlined and concentrated its air C2 structure, resources, and associated training efforts at two static air-operations center (AOC) locations (Torrejon, Spain; and Uedem, Germany) and in one deployable AOC headquartered in Poggio Renatico, Italy. This more focused effort has increased training proficiency and readiness levels across the board in support of current allied air operations.

NATO can also exercise tactical-level C2 closer to actual air operations. NATO operates 17 E-3A airborne warning and control system (AWACS) aircraft stationed at Geilenkirchen, Germany. These aircraft are part of the NATO Airborne Early Warning and Control (NAEW&C) Program established in 1978.

The United States, Belgium, the Czech Republic, Denmark, Germany, Greece, Hungary, Italy, Luxembourg, the Netherlands, Norway, Poland, Portugal, Romania, Spain, and Turkey participate as full members in NAEW&C. The United Kingdom's (UK) seven AWACS aircraft also participate in the program and would be a key part of any NAEW&C effort. With the historic return of France to NATO military operations in March 2009, its four E-3F aircraft could also now be available.[4]

The current allied air commander, General Frank Gorenc, recently highlighted the continuing value of the AWACS:

The NATO Airborne Early Warning and Control E-3A Force has been absolutely critical to the success of NATO operations and providing Air Battle Management, Command and Control and Situational Awareness for the Alliance. The versatility of the E-3A force continues to be demonstrated today as we have seen during the ongoing crisis in the Ukraine.[5]

AIRMEN

Air forces might have the best aircraft available, but without solid airmen of all ranks to operate and maintain them, they will not be successful. From general officers to midgrade and junior officers to the noncommissioned officer and enlisted force, NATO nations have air force leaders and airmen of great capacity. Senior leaders have a passionate vision of air power's importance to the security of the alliance and to any military campaign or operation.

However, these leaders are also struggling with how to provide the best airpower capability to their respective nations and NATO under a fiscally constrained environment. NATO nations are making huge strides in professionalizing and recognizing the value of the enlisted force. For example, Poland, a relatively new member of the alliance that has a military that once largely consisted of conscripts, has made significant investments in training and professionalizing its air arm.

AIRCRAFT

NATO member nations collectively have well over 3,000 tactical-fighter aircraft of various types and capabilities. Germany, Italy, Spain, and the UK all operate the modern Eurofighter Typhoon. France operates the fourth-generation Rafael and the latest versions of the Mirage. Several countries fly older F-16s, but ones with midlife upgrades— Denmark, the Netherlands, Norway, Portugal, and soon Romania.

Turkey flies newer Block 30, 40, and 50 F-16s, and Greece and Poland fly Block 52 F-16s. Spain and Canada operate very capable F-18s. Several countries still fly older but very capable aircraft, such as the Tornado, Mirage, and Phantom. Finally, some countries fly fleets of Russian-produced aircraft with Western-style, NATO-compatible modifications. For example, Poland recently deployed four MiG-29s to do Baltic air policing over Estonia, Latvia, and Lithuania.[6]

NATO Partnership for Peace (PfP) nations—Sweden, Finland, and Austria—also have advanced fighter capabilities. While there are more than enough aircraft available, the issue is whether nations will commit enough of these aircraft to fly specific types of missions in particular NATO operations. As former Deputy Assistant Secretary of Defense for Europe and NATO policy Ian Brzezinski has noted, "Little more than a handful of NATO's 28 members proved willing to fly strike missions in Libya."[7]

Areas for improvement in non-U.S. NATO aircraft are 1) stealth capability, 2) air-refueling tankers, 3) strategic airlifters, and 4) intelligence, surveillance, and reconnaissance (ISR) aircraft. NATO operations have long relied on the U.S. Air Force for the bulk of these types of aircraft. These challenges and non-U.S. NATO nations' efforts to address them will be discussed later.

MUNITIONS

Collectively, NATO nations have a variety of close-in, precision-guided weapons. They also have a number of precision standoff munitions that can suppress enemy air defenses and conduct standoff precision strikes so aircraft do not have to penetrate lethal air-defense rings. However, potential adversaries' air defenses continue to advance and mature, creating increased risk and difficulties for air forces to provide traditional close-in, air-to-ground weapons employment. As Brzezinski said:

> The preponderance of the initial salvo that disabled Gadhafi's air defense came from U.S. forces, and afterwards U.S. aircraft were relied on heavily for intelligence gathering, surveillance, air-to-air refueling, electronic jamming, and the suppression of enemy air defenses. European allies soon ran out of precision-guided munitions and other key wartime consumables and had to turn to U.S. inventories for replenishment.[8]

At a 2014 Air Force Association conference, General Philip Breedlove, NATO's Supreme Allied Commander Europe, also addressed the need to maintain larger NATO stockpiles of standoff and precision weapons:

> . . . what we learned in Libya and other places, we do not have enough precision strike munitions to carry on a concentrated campaign at length helping all of our allies to be there with us. I think we need to think through where we are on precision munitions.[9]

Likewise, upgrading these weapons' precision and survivability in response to evolving and improving enemy defense countertechnologies will be essential for any future operations.

BASING

Basing is a tremendous strength for NATO. It is superb across NATO and at the extreme edges of the alliance. Poland, Romania, Bulgaria, and Turkey provide excellent airfields in the far eastern portions. When Kyrgyzstan asked the United States to depart Manas Air Base by July 2014, Romania quickly volunteered

Mihail Kogălniceanu Air Base as the new multimode logistics location for flow in and out of Afghanistan.

Greece, Italy, France, and Spain also provide first-rate bases in the extreme southern boundaries, as seen during Operation UNIFIED PROTECTOR. Portugal provides basing in the far western boundary. Iceland, the UK, Denmark, Norway, Lithuania, and Estonia provide robust basing options for operations in the northern borders. The question, again (as with aircraft), is whether nations will make some or all of their bases available for a specific NATO operation or even restrict the type of aircraft and mission that can fly from a base.

Basing is a straightforward example of NATO's Smart Defence principle. Not every nation needs to have every aspect of airpower in its inventory. Estonia is a good example. The Estonian Air Force is small but has a great airfield, and it makes available to all NATO air forces. Likewise, Estonia provides some of the best cyber expertise to the alliance and hosts NATO's Cooperative Cyber Defence Centre of Excellence. In brief, Estonia provides the kind of pooled, shared, and coordinated capabilities that NATO needs in a time of budgetary austerity.[10]

AIR AND MISSILE DEFENSE

Many NATO nations have surface-based air-defense systems to defend against aircraft threats. Surface-based air-defense missiles and surveillance radars are spread among various service components in each country, but for any NATO operations, they would fall under the C2 of the Allied Air Command. In combination with advanced air defense, multi-role fighters, and AWACS, non-U.S. NATO nations can adequately defend airspace against enemy attack.

However, there is a shortfall in BMD capabilities. Both the 2010 and 2012 NATO summits identified the need to strengthen BMD. At the 2012 summit, NATO declared that it had an interim BMD capability based largely on U.S. contributions under the U.S. Missile Defense Agency's European Phased Adaptive Approach (EPAA).[11] The NATO interim capability consisted of the integration of a U.S. Navy Aegis ship; a U.S. land-based, long-range radar; and a C2 capability located at Allied Air Command.

A second phase of the EPAA, to be implemented in 2015, will make this BMD capability more robust through improvements to the Aegis radar and its defense missile capabilities. This phase will also add an Aegis-ashore system in Romania and upgrade the C2 systems.

In 2018, phase three will provide another round of Aegis software and missile updates, add a second Aegis-ashore system in Poland, and provide more advances in C2 systems at Allied Air Command. NATO leadership continues to encourage alliance nations to provide additional capabilities (or funding) to improve NATO coverage against ballistic missile threats, and NATO nations are responding, as captured in the 2014 Wales Summit Declaration:

> 58. Today we are also pleased to note that additional voluntary national contributions have been offered, and that several Allies are developing, including through multinational cooperation, or are acquiring further BMD capabilities that could become available to the Alliance. Our aim remains to provide the Alliance with a NATO operational BMD that can provide full coverage and protection for all NATO European populations, territory, and forces, based on voluntary national contributions, including nationally funded

interceptors and sensors, hosting arrangements, and on the expansion of the Active Layered Theatre Ballistic Missile Defense (ALTBMD) capability. Only the command and control systems of ALTBMD and their expansion to territorial defence are eligible for common funding.

59. We note the potential opportunities for cooperation on missile defence, and encourage Allies to explore possible additional voluntary national contributions, including through multinational cooperation, to provide relevant capabilities, as well as to use potential synergies in planning, development, procurement, and deployment. We also note that BMD features in two Smart Defence projects.[12]

With the Syrian civil war potentially spilling over into Turkey in 2012, Ankara requested support from its NATO allies to bolster its air defense and BMD capabilities. In response, On December 4, 2012, NATO foreign ministers agreed to the request, with Germany, the Netherlands, and the United States deploying a total of six Patriot-missile batteries in response. By February 16, 2013, the last battery had arrived and was operational. This deployment should be hailed as an example of successful collaboration and flexibility in NATO's growing BMD role.

Spanish Defense Minister Pedro Morenés announced in September 2014 Spain's intent to deploy Patriots to Turkey in January 2015. Spanish missiles and soldiers are expected to replace the two Dutch batteries deployed in Adana, Turkey.[13]

A near-term challenge for NATO air and missile defense, however, will be intertwining various defense systems. Currently, there is a mix of former Soviet and current western systems. Possibly adding to this problem, in September 2013, Turkey publicly announced that it intended to procure Chinese-made air-defense systems.[14] The United States and other NATO countries made their displeasure with that decision clear, and Congress took the unusual step of actually writing into the defense authorization bill a provision denying the use of any American government funds to help "integrate the Chinese missile defense systems into U.S. or NATO systems."[15] Ankara has since postponed its decision to procure the Chinese systems to consider purchasing alternative systems.[16]

READINESS

Regular NATO exercises are intended to hone NATO airmen's skills at both the operator and command-and-control levels. Likewise, individual NATO nations often host or participate in bilateral and multilateral exercises. For example, three to four times a year, Spain hosts the Tactical Leadership Program to train aircrews in large mission employment, and Portugal hosts the multilateral Real Thaw exercise every year. NATO nations also participate in air exercises outside of Europe. Israel's most recent Blue Flag exercise included aircraft and airmen from the United States, Italy, and Greece.

NATO's Allied Air Command recently conducted its largest exercise to date, Ramstein Ambition II 2014—a computer-assisted, command-post exercise simulating continuous operations—in which 400 airmen from 26 nations participated. According to General Gorenc, Ramstein Ambition II 2014 is a great validation point on the march toward AIRCOM's full operational capability.[17]

However, it is important to note that each nation funds its own participation in most exercises. If

NATO nations continue decreasing defense funding, the level of participation in NATO-level exercises and nationally hosted multilateral or bilateral training will almost certainly fall as well. To forestall this, NATO and alliance militaries should be examining increased use of linked, high-end simulators.

Simulation capabilities are advancing every year, enabling aircrews to make "fatal" mistakes that they learn from without experiencing the costly loss of life and aircraft resources. The challenge with linking high-fidelity aircraft simulators is that it requires a significant upfront investment from NATO nations.

CURRENT CHALLENGES

NATO's leaders currently face 10 key challenges to the alliance's air capabilities and operations. The following subsections detail NATO's current plans to address these issues and identify what it can do to overcome these shortages in a more effective and sustainable manner.

Stealth Capability.

Stealth capability is not a luxury; it is a necessity in the context of the advancing defense designs that NATO airmen could face. Currently, the United States provides the only stealth aircraft capability. But the F-35 Joint Strike Fighter (JSF) will fill this void for non-U.S. NATO nations, seven of which have ordered or stated intent to buy a total of 512 F-35s: the UK, the Netherlands, Italy, Canada, Denmark, Norway, and Turkey.[18] This international venture is a win for NATO. The fifth-generation F-35 provides a commonality of logistics and tactics for any future NATO operation and reduces the number of support-package aircraft required, such as those used for dedicated electronic attack or air superiority.

ISR.

The strategic level of ISR can be divided into space-based and aircraft-based systems. The larger NATO nations have their own space-based systems or have collaborated for many years to cooperatively fund, develop, launch, operate, and sustain various types of space surveillance systems that would support a NATO military effort.

Regarding aircraft, both the French and the British have strategic- and operational-level ISR aircraft. For example, the UK operates five Sentinel aircraft that have advanced surveillance radars mated to Bombardier Global Express business jets. The system reached initial operational capability in July 2008. The UK also just received the first of two Northrop Grumman E-8 Joint Surveillance Target Attack Radar System aircraft from the United States to replace their aging Nimrod R1.

Likewise, the Alliance Ground Surveillance (AGS) system is scheduled to reach initial operating status (with basing in Italy) by 2016. NATO AGS will use five remotely piloted Block 40 Global Hawks. Fifteen NATO member countries are currently contributing to the acquisition of the aircraft—Bulgaria, the Czech Republic, Denmark, Estonia, Germany, Italy, Latvia, Lithuania, Luxembourg, Norway, Poland, Romania, Slovakia, Slovenia, and the United States.[19] These high-in-demand, short-in-supply aircraft will undoubtedly be a great utility in almost any NATO contingency.

The UK and Italy provided remotely piloted aircraft (RPAs) for NATO operations in Afghanistan (MQ-9 and MQ-1 RPAs, respectively). Italy began accepting six total MQ-9s in 2011, and France accepted two of an eventual 12 MQ-9s in January 2014 in an effort to replace its less-capable Harfang RPA. Germany and the Netherlands have also expressed interest in operating tactical-type RPAs. The NATO alliance is growing its ISR capability and must maintain this momentum in light of the reduction of RPA systems in future U.S. defense budgets. ISR is one of the main focus areas for NATO's Smart Defence concept.

As NATO increases its ISR capability, it must also grow a parallel processing, exploitation, and dissemination capability for the data it gathers. The United States learned some hard lessons in this area when it rapidly grew its MQ-1, MQ-9, and Global Hawk force. The U.S. Air Force is sharing these lessons with NATO nations to help them avoid similar growing pains.

Investment Levels.

NATO has a long-established defense spending goal of 2 percent of each member nation's gross domestic product (GDP). Unfortunately, only four nations (the United States, the UK, Greece, and Estonia) achieved that goal in 2013, and many even decreased spending levels.[20] The general trend is in the wrong direction, but Poland is a notable exception: it has increased defense spending, up to 1.9 percent of GDP in 2013,[21] and discussions at the 2014 Wales summit indicated a commitment and intent by the nations to reverse the trend. An excerpt from the Wales summit states:

We agree to reverse the trend of declining defence budgets, to make the most effective use of our funds and to further a more balanced sharing of costs and responsibilities. Our overall security and defence depend both on how much we spend and how we spend it. Increased investments should be directed towards meeting our capability priorities, and Allies also need to display the political will to provide required capabilities and deploy forces when they are needed. A strong defence industry across the Alliance, including a stronger defence industry in Europe and greater defence industrial cooperation within Europe and across the Atlantic, remains essential for delivering the required capabilities. NATO and EU efforts to strengthen defence capabilities are complementary.[22]

In these austere times for most of Europe, nations are carefully balancing defense dollars across personnel, training, sustainment, current national operations, and future capabilities. Nations should consider making the same tough decisions the United States had to make in carefully cutting personnel to afford more military hardware.

Notably, when it comes time to deploy for a NATO operation, it is a pay-your-own-way system. Establishing a common fund for operations could encourage more national airpower contributions to any NATO operation. But creating a common operational fund is problematic, since NATO would either have to look for donors or tax each nation a percentage of its defense budget (or GDP). At this point, it may be easier to simply continue with the pay-your-own-way model.

National Caveats.

Once NATO becomes engaged in a particular operation, nations will sometimes have caveats (or particular restrictions) on the types of missions they will fly or how they will execute portions of missions. Though not insurmountable, this effectively handcuffs the NATO joint force air-component commander (JFACC) planning and execution. When allocating forces to a mission, the JFACC must consider these caveats and plan around them. Elimination of all mission caveats is the ultimate goal. In the absence of that, nations should provide NATO air-planning staffs with a list of their most likely national caveats well in advance of any operation.

In a June 2011 visit to Brussels, then–U.S. Secretary of Defense Robert Gates expressed his frustrations with operations in Afghanistan, noting that the war effort had been hobbled by "national 'caveats' that tied the hands of allied commanders in sometimes infuriating ways."[23] Ironically, the United States effectively imposed national caveats during Operation UNIFIED PROTECTOR in Libya by not allowing its air and naval air forces to perform strike missions; these forces only performed enabling missions such as ISR, air refueling, suppression of enemy air defenses, and electronic attack. For the first time, and despite Washington's previous complaints about allied caveats, the United States became, in the words of Brzezinski, a "caveat nation."[24]

Strategic Lift.

Strategic airlift is the key to moving troops, equipment, or a fighter or ISR aircraft package to the optimum location within or outside of NATO. However, among allied states and as seen in the French operation in Mali in 2013, this necessary capability is limited and currently relies too heavily on the U.S. C-17 and C-5 strategic airlift force.[25] Currently, the UK has eight C-17s that would be used extensively in any NATO operation.[26] Canada has four C-17s, and 10 NATO nations have access to a separate group of three C-17s, known as the Strategic Airlift Capability (SAC).

Established in September 2008 in Hungary, the Heavy Airlift Wing (HAW) conducts SAC operations. The HAW is not a NATO organization, but a number of NATO and PfP nations—including Hungary, Bulgaria, Estonia, Lithuania, the Netherlands, Norway, Poland, Romania, Slovenia, the United States, Finland, and Sweden—contribute personnel and money for access to a proportionate share of the HAW's annual flying hours.

Many nations also collaborate to meet their strategic airlift needs with commercial aircraft they contract from other non-NATO nations. However, these aircraft may not be available in a conflict, for a variety of reasons. The future looks better with the purchase of the Airbus A-400M strategic airlifter by Germany (53), France (50), Spain (27), Turkey (10), Belgium (7), and Luxembourg (1).[27] Deliveries began in 2013 and will extend through 2024.[28] The UK is purchasing 14 of the A-330 Multi-Role Tanker Transport (MRTT) aircraft, and France is hoping to purchase 12 MRTTs.[29]

Air Refueling.

The United States provided the majority of the air-refueling capability for NATO's 1999 Operation ALLIED FORCE in the Balkans. It also provided approximately 80 percent of all the air refueling missions in Operation UNIFIED PROTECTOR in 2011 and sent tankers in support of French fighters and bombers during France's 2013 Mali operation.

A March 2014 analysis by the Joint Air Power Competence Centre summarized NATO's current air-refueling capability with and without U.S. contributions. With U.S. capability, NATO has 709 air-tanker-capable aircraft spread across multiple aircraft types, some with boom-type capability, some with drogue-type capability, and some with both. Without U.S. capability, NATO nations collectively have only 71 air-refueling aircraft, many of which are aging and are spread across multiple types of airframes.[30] For example, the French operate three KC-135s and 11 C-135Rs, Turkey operates seven KC-135Rs, the UK operates four Lockheed TriStars, and the Netherlands operates two KDC-10s.[31]

The outlook is improving somewhat. In 2011, Italy received four new Boeing KC-767s with drogue and boom capability. Germany and Canada operate small fleets of modern military A-310 Airbus cargo and passenger aircraft with extra fuel capacity and a probe-and-drogue system added to each wing.

As noted previously, in the near term, the UK is procuring 14 Airbus A330 MRTT, with 9 of 14 in operation as of May 2014, while France announced that it intends to buy 12. In addition, some of the NATO nations buying the A-400M plan to equip them with underwing drogue-refueling systems. The challenge for NATO will be achieving the right mix of boom and drogue capability to match the NATO fighter aircraft fleets' current and future requirements.

However, if defense budgets remain strained and some planned procurements are put to the side, a remedial strategy could be for nations to create a shared-tanker capability similar to the C-17 HAW. Participating nations would contribute dollars and personnel to a common air-refueling capability and then get access based on their prorated contribution.

The Consensus Mechanism.

All NATO decisions are made by consensus after discussion and consultation that give alliance members the opportunity to exchange views and information. Certainly, this process can produce well-thought-out actions with thorough discussions of possible unintended second- or third-order consequences. But decisiveness is not its hallmark, and the challenge is reaction time.

Articles 4 and 5 of the NATO charter provide key principles for how the alliance consults and takes action. Article 4 effectively says that any nation can bring security issues and concerns to the North Atlantic Council for discussion and can seek NATO help in bolstering defense. Nations have invoked Article 4 only four times in NATO history; most recently, Poland invoked it after Russia invaded Crimea. The three previous times, Turkey invoked Article 4: in 2003 at the start of the Iraq War, in June 2012 after Syria shot down a Turkish military jet, and in October 2012 after Syrian attacks in Turkey.[32]

Article 5 is the basis of a fundamental principle of NATO: collective defense. The article provides that if a NATO ally is the victim of an armed attack, every other member of the alliance will consider this act of violence an armed attack against all members and will take the actions it deems necessary to assist the ally that is attacked. NATO has only invoked Article 5 once, following the September 11, 2001, terrorist attacks.[33]

Airpower is the most rapid response capability NATO has. The NATO Response Force, which includes an air component, is designed for crisis response and to do three things: deploy as a standalone force for Article 5 operations or non–Article 5 crisis response, deploy as an initial entry force until larger forces can arrive, and deploy as a demonstrative force to deter a crisis.[34] Yet, the NATO staffing and consensus process can be lengthy, even for Article 5. When NATO reaches consensus, nations must still offer force capability for the agreed-to operation. During this waiting period, individual NATO nations may take unilateral or multilateral action outside of NATO.

One example of the lengthiness of the consensus process is Libya. While NATO eventually supported UN Security Council Resolutions and led Operation UNIFIED PROTECTOR, the first missions did not occur until March 27, 2011, almost 10 days after the UN Security Council Resolutions were published and long after France started the initial strikes. By the time NATO took over, a coalition of NATO and non-NATO countries was already executing combat air operations.

Following Russian aggression in Ukraine, Poland sought Article 4 consultations on March 1, 2014. In response, NATO leaders met from March 2-4 to discuss possible actions but did not declare any additional defensive actions. Within a week, the United States had bilaterally deployed an additional six F-15C Eagle aircraft and a KC-135 tanker to bolster the ongoing U.S. rotation in the NATO Baltic air-policing mission. Likewise, by March 9, the United States had bilaterally increased the size of an already-planned F-16 exercise with Poland to 12 aircraft. Within 2 weeks, NATO was flying surveillance missions over alliance territory in the proximity of Ukraine, using its NAEW&C program E-3As.

During an April 3 press engagement at NATO headquarters with the new Estonian prime minister, the NATO secretary general said:

> We have more than doubled the number of fighter aircraft allocated to our air policing mission in the Baltic States, thanks to the United States. Many European Allies have also offered additional planes, air-to-air refueling tankers and other capabilities. And we will make sure that we have updated military plans, enhanced exercises and appropriate deployments.[35]

However, it was not until after a North Atlantic Council meeting on April 16, that the secretary general formally announced larger NATO air, land, and sea responses to bolster the defense of the Baltics and Poland.[36] By early May 2014, NATO was deploying these non-U.S. aircraft to Poland and to the Baltic air-policing mission (replacing the bolstered U.S. F-15 rotation). As General Gorenc noted:

> What you see here is Allied solidarity. Under our long-standing plans for NATO's Baltic Air Policing, the Polish Air Force deployed MiG-29 fighters in May, leading the mission from Siauliai Air Base, Lithuania. The effort has been supported by Royal Air Force Typhoons also flying out of Siauliai and Royal Danish Air Force F-16s flying out of Amari, Estonia. At the same time France has supported the mission with its Mirage 2000 fighters here at Malbork (Poland).[37]

Long-Range Bombers.

Non-U.S. NATO air forces do not have a long-range bomber capability, despite the efficacy of such a platform: it can operate, without needing refueling, at long ranges and with heavy precision payloads and long target loiter times. Because of declining budgets and NATO nations' closer proximity to their likely areas of operation, it is probably beyond the scope of any single NATO nation to procure such a capability. But could NATO nations agree to a commonly funded long-range bomber capability using a model such as the NAEW&C, NATO AGS, or even the non-NATO HAW? At a minimum, it is important that NATO maintain bases capable of hosting forward-deployed U.S. bombers. As the United States designs and procures a new long-range bomber, it must consult closely with NATO allies to ensure that some existing and future NATO airfields can host the aircraft.

A Common Vision of Strategic Threats.

Is the resurgence of Russia in the East or the terrorist threats emanating from Africa and the Middle East the main strategic threat to NATO? Arguably, both are. Therefore, non-U.S. NATO air forces must train for both high-end and counterinsurgency-type conflict. Strategic and tactical ISR platforms are crucial to both efforts, so non-U.S. NATO nations must maintain ISR investment strategies for the future. The Chicago NATO summit in May 2012 reinforced the need to strengthen multinational cooperation—in particular, on some strategic programs, including the AGS program. At a press briefing on March 5, 2012, NATO Secretary General Anders Fogh Rasmussen said:

> We will target a number of strategic projects for 2020 and beyond. As our operation in Libya showed, we still face some specific capability gaps, such as air-to-air refueling and joint intelligence, surveillance and reconnaissance. And we know that we will need stron-

ger cooperation, across the Atlantic and in Europe, to fill them.[38]

Cyber Preparedness.

Integrating cyber readiness into air operations is absolutely critical for NATO to keep positive C2 of assets and missions. Admiral James Stavridis, former supreme allied commander Europe, commented on cyber preparedness when he was commander:

> Top of my list. Here we see the greatest mismatch between the level of potential threat and our preparation for it. While the 28 NATO nations collectively have enormous skill and capability in this area, we have yet to find ways to work together, largely due to national caveats and concerns about sharing such sensitive technology, intelligence, and knowledge.[39]

Last November, NATO kicked off its annual Cyber Coalition exercise in Estonia. Jamie Shea, NATO's deputy assistant secretary general for emerging security challenges, explained:

> Cyber-attacks are a daily reality, and they are growing in sophistication and complexity. NATO has to keep pace with this evolving threat and Cyber Coalition 2013 will allow us to fully test our systems and procedures to effectively defend our networks — today and in the future. . . . NATO has to keep pace with this evolving threat.[40]

Cyber was the focus topic at the November 2013 International Seminar of the Alfredo Kindelán Chair — a renowned forum for the study and debate of military air strategy and doctrine — in Madrid.[41] The conference's keynote speaker discussed cyber preparedness in air operations and challenged NATO air force leaders in the audience to consider:

1. If I suffer a cyber attack, do I know? Do active, layered network defense sensors alert me?

2. Once I realize I am under attack, do I have a reporting procedure and repair plan that isolates the attack and gracefully degrades air C2 to a backup plan (if required)?

3. Am I truly prepared? Have I practiced 1 and 2 above?[42]

Since then, NATO has updated its cyber defense policy. The new policy considers a cyber attack no differently than an attack with conventional weapons, stating that cyber attacks are covered by Article 5. The new cyber policy was approved by defense ministers and gained endorsement at the 2014 NATO summit.[43] This is the warfare of the future, and NATO and its airmen are preparing for it.

CONCLUSION

NATO C2, airmen, aircraft, munitions, basing, air and missile defense, and readiness are all pertinent factors when examining the status of NATO air power without a U.S. capability. NATO is addressing each of the 10 challenges outlined earlier, but it is doing so with budgets that may or may not allow it the resources to fully fix these shortages.

Under NATO's Smart Defence banner, coordination among nations to procure similar capabilities will be key. Nations should consider where they have expertise and capability to contribute and should not procure unnecessary, duplicative capabilities that other NATO nations could provide. The NAEW&C

and AGS programs should be considered the norm for the future. Pooling resources to share airlift or tankers with organizational construct like the HAW in Hungary could be essential to the future success of non-U.S. NATO air power.

During the U.S. Air Force Association's September 2011 conference, French Air Force General Stéphane Abrial, former NATO supreme allied commander for transformation, said that non-U.S. NATO air forces "could not have performed to the same level of effectiveness without heavy contribution from the U.S." and would be severely limited if the United States chose not to join a foreign operation such as the one conducted in Libya.[44]

Although it is doubtful that NATO would ever participate in a major conflict without significant U.S. contributions, the fact is that in the 1990s, the U.S. Air Force presence in Europe was much larger than it is today. Numbers do count in any conflict. Non-U.S. NATO nations must maintain their current air force capabilities while procuring more advanced capabilities, such as the A-330 MRTT, A-400M, and JSF. They must also procure enough advanced standoff munitions for any projected conflict.

During fierce internal budget battles, vocal ministers of defense will be key to NATO's goal of each nation spending 2 percent of its GDP on defense. Certainly, recent Russian aggression in Ukraine and the brutality of the Islamic State in Iraq and Syria should provide a wake-up call to NATO's national capitals. The climate could be the necessary impetus to spend more on defense and, in turn, commit forces to future NATO endeavors.

ENDNOTES - CHAPTER 14

1. This chapter was originally published as an essay on October 30, 2014.

2. NATO, "Smart Defence," Brussels, Belgium, NATO, July 16, 2014, available from *www.nato.int/cps/en/natolive/topics_84268.htm*.

3. Headquarters: Allied Air Command, "AIRCOM Focus Areas," Ramstein Air Base, Germany, available from *www.airn.nato.int/04FocusAreas/01focusareas.html*.

4. NATO NAEW&C Programme Management Agency, "NAEW&C Programme Management Organisation,"Brunssum, The Netherlands, available from *www.napma.nato.int/organisation/2.html*.

5. Headquarters: Allied Air Command, "AIRCOM Leadership at the E-3A Component," Ramstein Air Base, Germany, March 24, 2014, available from *www.airn.nato.int/03NewsRoom/022014/1405.html*.

6. NATO, "Allies Enhance NATO Air-Policing Duties in Baltic States, Poland, Romania," Brussels, Belgium, April 30, 2014, available from *www.nato.int/cps/en/natolive/news_109354.htm*.

7. Ian Brzezinski, "Lesson from Libya: NATO Alliance Remains Relevant," *National Defense*, November 2011, available from *www.nationaldefensemagazine.org/archive/2011/November/Pages/LessonFromLibyaNATOAllianceRemainsRelevant.aspx*.

8. *Ibid.*

9. Air Force Association, Annual Air & Space Conference and Technology Exposition 2014: The Future of NATO (transcript, September 15, 2014, National Harbor, Maryland, available from *www.airforcemag.com/afatranscripts/Documents/2014/September%20 2014/091514breedlove.pdf.pdf*.

10. "Smart Defence," *NATO Review*, available from *www.nato.int/docu/review/Topics/EN/Smart-Defence.htm*.

11. NATO, NATO Ballistic Missile Defence, October 2012, available from *www.nato.int/nato_static/assets/pdf/pdf_2012_10/20121008_media-backgrounder_Missile-Defence_en.pdf.*

12. NATO, Wales Summit Declaration, September 5, 2014, available from *www.nato.int/cps/en/natohq/official_texts_112964.htm.*

13. NATO, Allied Command Operations, "General Breedlove Thanks Spain for Patriot Commitment," news release, September 18, 2014, available from *aco.nato.int/general-breedlove-thanks-spain-for-patriot-commitment.aspx.*

14. Edward Wong and Nicola Clark, "China's Arms Industry Makes Global Inroads," *The New York Times*, October 20, 2013.

15. Denise Der, "Why Turkey May Not Buy Chinese Missile System After All," *The Diplomat*, May 7, 2014, available from *thediplomat.com/2014/05/why-turkey-may-not-buy-chinese-missile-systems-after-all/.*

16. "Made in China? U.S. Warns Turkey Its Missile Deal with Beijing May Be Incompatible with NATO," *RT*, October 24, 2013, available from *rt.com/news/us-turkey-china-missile-defense-689/.*

17. Headquarters: Allied Air Command, "An Ambitious Exercise at AIRCOM," Ramstein Air Base, Germany, available from *www.airn.nato.int/03NewsRoom/022014/1409.html.*

18. Lockheed Martin, "F-35 Lightning II Program Status and Fast Facts," Bethesda, MD, December 8, 2014, available from *www.lockheedmartin.com/content/dam/lockheed/data/australia/documents/F-35FastFactsDecember2014.pdf.*

19. Northrop Grumman, "Photo Release—Northrop Grumman Begins On-Time Production of First NATO Global Hawk," Mechanicsburg, PA, December 3, 2013, available from *www.globenewswire.com/newsarchive/noc/press/pages/news_releases.html?d=10060279.*

20. Ewen MacAskill, "U.S. Presses NATO Members to Increase Defence Spending," *The Guardian*, June 23, 2014, available

from *www.theguardian.com/world/2014/jun/23/us-nato-members-increase-defence-spending.*

21. "10 Facts: Sizing Up NATO's Defense Spending," *The Globalist,* March 30, 2014, available from *www.theglobalist.com/10-facts-sizing-up-natos-defense-spending/.*

22. NATO, Wales Summit Declaration, para. 14.

23. Thom Shanker, "Defense Secretary Warns NATO of 'Dim' Future," *The New York Times,* June 10, 2011.

24. Brzezinski.

25. United States Africa Command, "U.S. Airlift of French Forces to Mali," Kelley Barracks, Stuttgart-Moehringen, Germany, January 24, 2013, available from *www.africom.mil/Newsroom/Article/10206/us-airlift-of-french-forces-to-mali.*

26. "UK Shows Interest in Buying Another C-17," *Defense News,* November 24, 2013, available from *www.defensenews.com/article/20131124/DEFREG01/311240001/UK-Shows-Interest-Buying-Another-C-17.*

27. "A400M (Future Large Aircraft) Military Transport Aircraft," *airforce-technology.com,* available from *www.airforce-technology.com/projects/fla/.*

28. Organisation for Joint Armament Cooperation, "A400M—A Tactical and Strategic Airlifter," Bonn, Germany, available from *www.occar.int/340.*

29. "A330-200 MRTT Future Strategic Tanker Aircraft (FSTA), United Kingdom," *airforce-technology.com,* available from *www.airforce-technology.com/projects/a330_200/.*

30. Joint Air Power Competence Centre, *Air-to-Air Refuelling Consolidation—An Update,* Kalkar, Germany, March 2014, available from *www.japcc.org/wp-content/uploads/AAR-Consolidation_web.pdf.*

31. *The Military Balance 2014*, London, UK: International Institute for Strategic Studies, February 2014.

32. NATO, "The Consultation Process and Article 4," Brussels, Belgium, March 4, 2014, available from *www.nato.int/cps/en/ natohq/topics_49187.htm?selectedLocale=en*.

33. NATO, "Collective Defence," Brussels, Belgium, June 2, 2014, available from *www.nato.int/cps/en/natohq/topics_110496. htm?selectedLocale=en*.

34. Pavolka Dalibor, *What Is NATO and EU Response Force Good For?* Berlin, Germany: Centre for European and North Atlantic Affairs, available from *cenaa.org/analysis/what-is-nato-and-eu-response-force-good-for/*.

35. NATO, "Joint Press Point with NATO Secretary General Anders Fogh Rasmussen and the Prime Minister of Estonia, Taavi Rõivas," Brussels, Belgium, April 3, 2013, available from *www. nato.int/cps/en/natolive/opinions_108844.htm?selectedLocale=en*.

36. NATO, "Doorstep Statement by NATO Secretary General Anders Fogh Rasmussen Following the Meeting of the North Atlantic Council," Brussels, Belgium, April 16, 2013, available from *www.nato.int/cps/en/natolive/opinions_109231. htm?selectedLocale=en*.

37. Headquarters: Allied Air Command, "NATO Measures to Enhance Collective Security," Ramstein Air Base, Germany, available from *www.airn.nato.int/03NewsRoom/022014/1408.html*.

38. NATO, "Monthly Press Briefing," Brussels, Belgium, March 2012, available from *www.nato.int/cps/en/natolive/7169_84 865.htm?selectedLocale=fr*.

39. James Stavridis, "SACEUR: Europe's Defense Challenges," Washington, DC: Atlantic Council, April 22, 2013, available from *www.atlanticcouncil.org/blogs/natosource/saceur-europes-defense-challenges*.

40. NATO, "NATO Holds Annual Cyber Defence Exercise," Brussels, Belgium, November 26, 2013, available from *www.nato. int/cps/en/natolive/news_105205.htm*.

41. Joint Air Power Competence Centre, "Transforming Joint Air Power," *The Journal of the JAPCC*, No. 13, 2011, available from *www.japcc.org/wp-content/uploads/Journal_Ed-13_web.pdf*.

42. This information derives from a summary of the author's keynote speech at the 2013 International Seminar of the Alfredo Kindelán Chair in November 2013.

43. Steve Ranger, "NATO Updates Policy: Offers Members Article 5 Protection against Cyber Attacks," Washington, DC: Atlantic Council, June 30, 2014, available from *www.atlanticcouncil. org/blogs/natosource/nato-updates-policy-offers-members-article-5-pro- tection-against-cyber-attacks*; and NATO, Wales Summit Declaration, para. 72.

44. John A. Tirpak, "Lessons from Libya," *Air Force Magazine*, December 2011.

ABOUT THE CONTRIBUTORS

BRUCE E. BECHTOL, JR., is an associate professor of political science at Angelo State University.

PAUL CORNISH is professor of strategic studies in the Strategy & Security Institute at the University of Exeter.

DOROTHÉE FOUCHAUX is a French defense analyst who has worked at the European Society of Strategic Intelligence and the French Ministry of Defence, Delegation for Strategic Affairs, as a research officer.

CRAIG FRANKLIN served in the U.S. Air Force for more than 3 decades before retiring in April 2014. His last position was commander of the 3rd Air Force and 17th Expeditionary Air Force, Ramstein Air Base, Germany. Before that, he was vice director of the Joint Staff at the Pentagon.

PATRICK KELLER is the coordinator of foreign and security policy at the Konrad Adenauer Foundation in Berlin, Germany.

GUILLAUME LASCONJARIAS is a research adviser at the NATO Defense College.

MICHAEL MAZZA is a research fellow in foreign and defense policy studies at AEI.

BRYAN MCGRATH is the managing director of the FerryBridge Group (a defense consultancy) and is a former officer of the U.S. Navy.

ANDREW A. MICHTA is the M. W. Buckman Professor of International Studies at Rhodes College and a senior fellow at the Center for European Policy Analysis in Washington, DC.

GARY J. SCHMITT is co-director of the Marilyn Ware Center for Security Studies, the American Enterprise Institute for Public Policy, Washington, DC.

ANDREW SHEARER is a senior official in the State Government of Victoria's Department of Premier and Cabinet.

W. BRUCE WEINROD served as the U.S. Secretary of Defense's representative for Europe and as defense adviser to the U.S. mission at NATO from 2007 to 2009. He was Deputy Assistant Secretary of Defense for European and NATO policy from 1989 to 1993.

TOSHI YOSHIHARA is the John A. van Beuren Chair of Asia-Pacific Studies at the U.S. Naval War College.